SPECIFICATIONS OF PHOTOVOLTAIC PUMPING SYSTEMS IN AGRICULTURE

Sizing, Fuzzy Energy Management and Economic Sensitivity Analysis

SPECIFICATIONS OF PHOTOVOLTAIC PUMPING SYSTEMS IN AGRICULTURE

Sizing, Fuzzy Energy Management and Economic Sensitivity Analysis

IMENE YAHYAOUI
Federal University of Espírito Santo, Brazil

ELSEVIER
AMSTERDAM • BOSTON • HEIDELBERG • LONDON
NEW YORK • OXFORD • PARIS • SAN DIEGO
SAN FRANCISCO • SINGAPORE • SYDNEY • TOKYO

Elsevier
Radarweg 29, PO Box 211, 1000 AE Amsterdam, Netherlands
The Boulevard, Langford Lane, Kidlington, Oxford OX5 1GB, United Kingdom
50 Hampshire Street, 5th Floor, Cambridge, MA 02139, United States

British Library Cataloguing-in-Publication Data
A catalogue record for this book is available from the British Library

Library of Congress Cataloging-in-Publication Data
A catalog record for this book is available from the Library of Congress

ISBN: 978-0-12-812039-2

For Information on all Elsevier publications
visit our website at https://www.elsevier.com

Publisher: Joe Hayton
Acquisition Editor: Lisa Reading
Editorial Project Manager: Maria Convey
Production Project Manager: Mohana Natarajan
Designer: Victoria Pearson

Typeset by MPS Limited, Chennai, India

For my lovely parents: Saïda and Lamine

Life is not easy for any of us. But what of that?
We must have perseverance and above all confidence in our selves.
We must believe that we are gifted for something,
and that this thing, at whatever cost, must be attained.

Marie Curie

CONTENTS

FOREWORD

Nowadays photovoltaic (PV) systems, through their increasing flexibility and reliability in use, offer a unique and cost-effective solution to provide energy services to remote rural areas in the sectors of health care, education, communication, agriculture, lighting, and water supply. With the continuously and dramatically decreasing price of PV systems, growing experience is gained with the use of PV in agriculture and other productive activities, which can have a significant impact on rural development. However, there is still a lack of information on sizing and operating of such systems in common applications in the agriculture fields, such as, for instance, the supplying of pumps for irrigations purposes.

Imene's book focuses on sizing and energy management of a stand-alone PV system for tomatoes irrigation. She starts from a study case (a pumping system for a tomato field in the Northern Tunisia), then generalizing the conclusions in such a way to provide useful information to professional, academic, students, and users that are involved at different levels and ways in such type of systems. The author studies in depth the system performance when facing variations in climatic, geographic, and economic parameters. I was given the opportunity to meet the author, during her stage at the University of Catania, and I appreciated her enthusiasm and competence in approaching this subject from experimental point of view as well. Such characteristics are mirrored in this book, which appears also well-structured and informative.

Giuseppe Marco Tina, Professor at the University of Catania, Italy

This book concentrates on the sizing and energy management of autonomous photovoltaic (PV) installations for irrigation in well-sunned sites. The book is an interesting read for professionals and academics who work with PV systems, since, using a case study, the author offers simplified and practical explanations about how it is possible to optimize the size and manage the energy generated of the PV components to fulfill the water needs of the crops. The author concluded the book by an interesting analysis of the variations in economic and geographic parameters impacts on the viability of PV water pumping systems. Imene is a promising young researcher who has a good publication record in PV energy in top journals and good international contacts, which helped her to develop the book's content and to perform high quality future works.

Fernando Tadeo, Professor at the University of Valladolid, Spain

ACKNOWLEDGMENTS

Well, emotionally, I am writing here my dedications for my first book, which has been written with much love and admire from my part and with the support of many persons, who helped me and encouraged me, generously. Foremost, I dedicate it for my lovely parents, my sister and my brothers. Without your support, this work could not be possible to be achieved.

I warmly thank my PhD supervisor: Professor Fernando Tadeo (University of Valladolid, Spain), who still, encouraging me to fulfill my dreams in scientific research, one of them, publishing this book, the fruit of my PhD research work. His guidance helped me in all the time of research, and he still is my reference in self-confidence, perseverance, hardwork, and organization.

My gratitude and special thanks go to Professor Giuseppe Marco Tina (University of Catania, Italy) for his generous help and continuous constructive remarks and critics. I thank him and Pr. Fernando for providing me the book's forewords.

I warmly thank my thesis codirector, Professor Maher Chaabene (University of Sfax, Tunisia). He is my reference in renewable energy, and thanks to him, I discovered this research field. During my research, his critics and remarks were very constructive and helpful.

Many thanks to the National Meteorological Institute of Tunisia and the Agriculture Administry of Medjez El Beb, Tunisia, for providing the work with climatic and agricultural data of Medjez El Beb, Tunisia. I also thank the Spanish Ministry of Economy and Competitiveness (Minneco) for believing in my work and supporting it with the FPI grant: BES-2011-047807 and the project DPI2010-21589-C-05.

I thank all my colleagues and professors from primary school in Bejà, Tunisia, until my postdoc research at the Electrical Engineering Department of the Federal University of Espiritu Santo, Brazil. Special thanks to Dr. Rachid Ghraizi for his support and help.

PROLOGUE

Over the last few decades, photovoltaic energy has become an effective source to produce electricity that will be used either in isolated sites or injected into the grid. In isolated areas in particular, photovoltaic installations are already being used for pumping water for agriculture or human purposes, since photovoltaic installations are easy to install and, after installation, the maintenance cost is low. However, the inherent variability of the sources means that the installation has to be carefully sized, so as to provide an adequate energy management algorithm and to minimize the system cost.

So, this book focuses on the sizing and energy management of an autonomous photovoltaic installation used to pump water for irrigation in an isolated agricultural site. Then, an economic study for photovoltaic systems feasibility for such water pumping plants is detailed in depth and compared with systems that include diesel generator, in three different countries. Typically, photovoltaic installations are widely used in arid and semiarid regions, such as the Maghreb and the South of Europe, where in addition, there is an important availability of solar radiation. The correct operation of these installations is needed, not only to fulfill the water demand, but also to guarantee the optimum use of the photovoltaic energy and extend the life of the system components. These objectives can be ensured by a good sizing of the components and an optimum energy management, which will be detailed in depth in this book.

In fact, the first part of this book deals with a state of the art of the renewable energies sources. More precisely, this part focus on photovoltaic energy use in irrigation and tomatoes irrigation characteristics. The second part of this book is concerned with the system components modeling, in which some of them are experimentally validated. In addition, some techniques related to the maximum photovoltaic power extraction have been studied. Then, the components sizing of the photovoltaic irrigation installation, namely, the photovoltaic panels and the battery bank, are presented in the third chapter. Hence, an algorithm for the optimum sizing of the installation components has been established, based on the crops water requirements, the site climatic characteristics, and the restrictions inherent to the components. Then, the sizing algorithm has been validated using measured data of the target area (Medjez El Beb, Northern Tunisia).

The fourth part of the book deals with the energy management of the photovoltaic irrigation installation. Hence, a fuzzy logic based algorithm has been established, to manage the energy generated by the panels and stored in the battery bank. Fuzzy logic has been used, since it is easy to implement and the present study is based on the user knowledge. The main idea behind the algorithm is as follows: depending on the photovoltaic power generated, the battery depth of discharge, the water level in the reservoir, and the water flux, the connection and disconnection of the components is deduced by using some proposed fuzzy rules. The algorithm efficiency has been firstly evaluated by simulation and validated secondly by experiments, in a plant installed in the laboratory, with satisfactory results. In Chapter 5, Comparative Economic Viability of PV, Diesel, and Grid Connected Plants for Water Pumping: Sensitivity Analysis According to Geographical Parameters, a sensitivity analysis of the water pumping systems costs is detailed: the costs of the systems are compared in three countries: Tunisia, Spain, and Qatar.

Hence, this book presents, in a simplified method, the concept of using photovoltaic/batteries systems in supplying the water pumps, in such a way it gives to the reader, a satisfactory and complete explanation about the components sizing, the energy management, and the economic feasibility of photovoltaic irrigation installations.

CHAPTER 1

Renewable Energies and Irrigation

1.1 INTRODUCTION

The production of electric energy using fossil fuels (oil, coal, natural gas, etc.) has traditionally provided adequate costs, but produce greenhouse gases. In fact, fossil power generation is responsible for 40% of global CO_2 emissions (Belakhal, 2010). Nuclear power, which does not produce directly carbon dioxide, generally suffers from poor acceptance because of significant risks and costly waste storage (Chapman & Mc Kinley, 1987; Lutze & Ewing, 1988).

In this context, renewable energies are positioned as a solution to fossil fuel depletion (Henrik, 2007; Ibrahim, 2000). For remote sites, where the grid is not available, renewable energies provide an excellent solution, since the energy sources are abundant (namely, solar radiation and wind). Moreover, given the adequate attributed support, renewable energies can meet much of the growing demand at lower prices than those usually forecast for conventional energy (by the middle of the 21st century, renewable energy sources could account for three-fifths of the world electricity market) (Johansson, Kelly, Amulya, & Williams, 1992). Moreover, the electricity can be produced near the place of consumption and without producing greenhouse gases. Hence, autonomous installations based on renewable energies are used for different applications in remote sites.

For agricultural applications, the use of Renewable Energies (Kumar & Kandpal, 2007) is a promising solution, especially for remote sites. In fact, much research has studied the efficiency of renewable energies in agriculture and other critical sectors for developing countries, as in the target country, Tunisia (Purohit & Kandpal, 2005). Modern cultivation techniques require regular irrigation; especially in arid and semi-arid climates (Sumaila, Teh, Watson, Tyedmers, & Pauly, 2008), for which, farmers generally use diesel engine water pumps. Although this solution was efficient in the past, the continuous increase in fuel prices and the requirement that the user be present are the main disadvantages of these installations (IEA-AIE, 2013). Hence, renewable energies are considered a good solution for farmers without easy access to fuel or for remote sites, as in the case study. In fact, in this book, a specific implementation for a 10 ha of land in Medjez El Beb (latitude: 36.39°, longitude: 9.6°) planted with tomatoes is focused on. To irrigate the crops, a diesel engine is currently

Specifications of Photovoltaic Pumping Systems in Agriculture
DOI: http://dx.doi.org/10.1016/B978-0-12-812039-2.00001-6

being used, which is complicated, since the site is isolated. Given the significant solar radiation during the growing season in this site, the solution proposed here is to use a photovoltaic installation for water pumping.

The rest of the chapter is organized as follows: Section 1.2 gives some general ideas about the state of the art in renewable energies. Section 1.3 presents the use of renewable energies for irrigation. In this section, some irrigation plants configurations are explained. The tomatoes irrigation specifications are detailed in Section 1.4, in which some factors of planting tomatoes, namely, the soil, the climate, and the crop are explained.

1.2 RENEWABLE ENERGIES

Nowadays, most of the electric energy is obtained from fossil fuels like oil, coal, and natural gas or from nuclear energy (Ben Ammar, 2011). Given the growing need for energy and the reduction in fossil sources, new energy resources are required to meet global energy needs. Renewable energies, such as photovoltaic, solar thermal, wind, hydro, waves, and biomass, are the best placed to fill this gap (Chaabene, 2009). In fact, these clean energy sources are inexhaustible. These sources are currently interesting, thanks to the great technological progress, the huge investments for the development of the energy production systems, and the rapid growth in the use of renewable sources (Ben Ammar, 2011; Nema, Nema, & Rangnekar, 2009).

Renewable energies can be consumed directly by loads, transported to the distribution system or stored in storage components (Nema et al., 2009). In fact, in grid-connected areas, they can be used as supplementary sources.

In isolated installations, where autonomy is required, the difference between the renewable energies produced and the energy needed by the loads requires the diversification of the sources or the use of storage components, such as batteries (Belakhal, 2010). Indeed, other technologies such as fuel cells could be used to store energy (Chaabene, 2009), although, they are not yet profitable due to their complexity. Hence, the majority of off-grid installations use lead–acid batteries.

Renewable energies production depends on the site and weather conditions. For instance, solar panels are effective if they are installed in well sunlit areas. Similarly, wind turbines are installed in regularly windy places (Chaabene, 2009). The energy conversion can be classified into three main categories: electric (photovoltaic panels), thermal (solar thermal, geothermal, etc.), and mechanical (wind). Here, a brief description of the main energy sources currently in use is given (of course hybrid solutions have also been developed but they are outside the scope of this book (Bernal-Agustín & Dufo-López, 2009; Deshmukh & Deshmukh, 2008; Green, 1982)).

1.2.1 Photovoltaic energy

Each year, the Earth surface receives 1.79×10^8 kWh, which is equivalent to a continuous power of 1.729×10^{17} W. It has been evaluated that 23% of this energy is reflected directly back into space, 29% is absorbed in the atmosphere and converted to heat radiated within the infrared spectrum, with the remaining 48% of the energy supplying the hydrological cycles and photosynthesis (Nema et al., 2009). Taking into account the alternations of day and night and cloudy periods, the peak power is estimated to be 1 kW (Nema et al., 2009).

In photovoltaic (PV) panels, the solar energy is converted to electricity by the junction charge carrier (contact between two different types of semiconductors: p-type and n-type) (Chenni, Messaoud Makhlouf, Kerbache, & Bouzid, 2007; Yotaka, Kazuo, & Hitomi, 1984) (Appendix A). Although the material required for making the photovoltaic modules (Silicon) is abundant and inexpensive, the complexity of the construction techniques makes these modules relatively expensive (Xiao & Dunford, 2004). However, technological advances are being made to enhance their competitiveness, such as the use of maximum power point tracking (MPPT) techniques (de Brito, Galotto, Sampaio, de Azevedo Melo, & Canesin, 2013; Price, Stuart, Yang, Zhou, & You, 2011) and cells yield enhancement (EPIA, 2014), which makes them a good solution, especially for isolated areas, thanks to their simplicity in installation. These facilities increase the worldwide photovoltaic energy use (Gules, De Pellegrin Pacheco, Hey, & Imhoff, 2008).

The components configuration in the photovoltaic installations depends on the application. Some of them are detailed below.

1.2.1.1 Serial configuration

In this case, all the photovoltaic energy produced passes through the battery bank, is converted from DC to AC by the inverter, and then transferred to the AC load (Fig. 1.1) (Chaabene, 2009; Wang & Zhang, 2010).

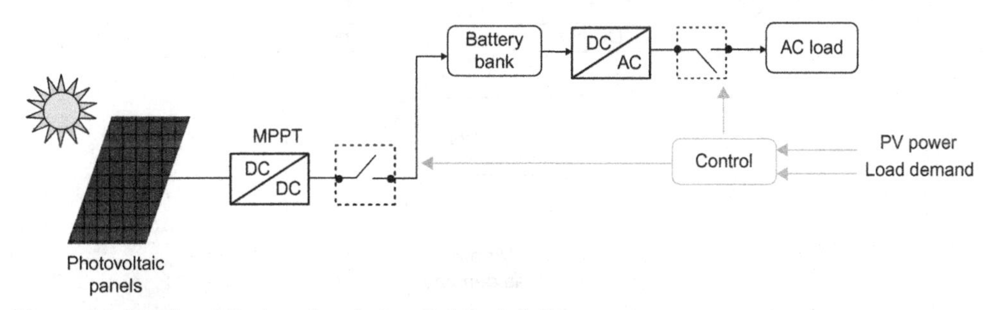

Figure 1.1 Serial architecture for photovoltaic installations.

This configuration is easy to install and can supply the load continuously. However, the excessive use of the battery bank decreases its lifetime. Moreover, it requires a large capacity to reduce its depth of discharge. Furthermore, the installation efficiency is reduced, since all the energy flows through the battery bank and the inverter.

1.2.1.2 Parallel configuration

The parallel configuration allows all energy sources to supply the AC load separately (Goswami, Kreith, & Kreider, 2000; Wang & Zhang, 2010) (Fig. 1.2).

In the case of an excess in photovoltaic energy generation, the bidirectional converter charges the battery bank. Hence, the load can be met by the PV panel, the battery bank, or both. Moreover, a reduction in the rated battery bank capacities, inverter, and photovoltaic panel is feasible, while also meeting the demanded load peaks and ensuring the installation autonomy and the battery efficiency (Goswami et al., 2000).

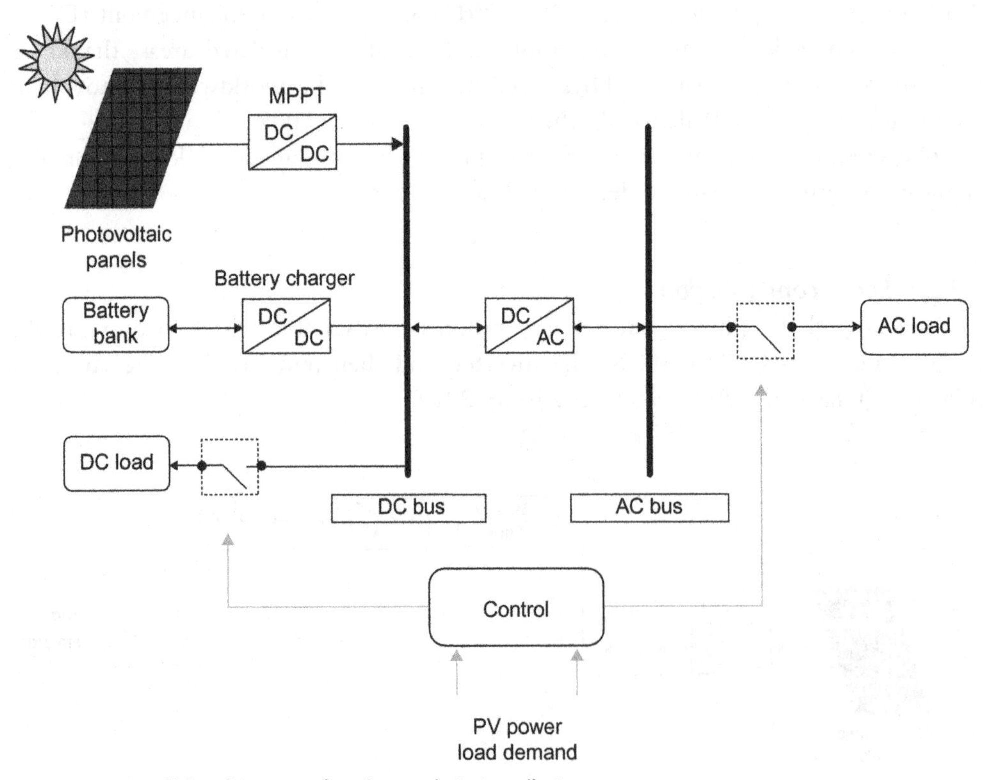

Figure 1.2 Parallel architecture for photovoltaic installations.

These objectives can only be met if the installed components are controlled by an "intelligent" energy management system. In fact, parallel systems include sophisticated controllers that include some of the following functions (Chaabene, 2009):

- Control of the energy flow based on the load energy demand.
- Battery low voltage disconnection, to prevent excessive discharging.
- Battery charging control that ensures fast recharge, while avoiding overcharge.
- Controlled "boost-charging" of flooded electrolyte lead–acid batteries at regular intervals (2–6 weeks) to reduce the negative effects of electrolyte stratification.
- Battery energy management based on voltage measurements to estimate the batteries state of charge.
- Controlled bidirectional energy flow through the inverter to allow the load to be supplied, and to charge the battery bank from renewable resources, when excess energy is available from the photovoltaic panel, which is operated at its maximum efficiency.

1.2.2 Thermal energy

Thermal energy consists of the use of heat to produce electricity. The most popular sources are the thermal panels and concentrators (solar thermal) or the high temperature from the Earth (geothermal) (Chaabene, 2009).

1.2.2.1 Solar thermal energy

The thermal conversion in this case consists in absorbing solar energy to heat up dark surfaces placed in sunshine. Solar energy collectors working on this principle consist of sun-facing surfaces which transfer part of the energy absorbed to a fluid (http://outilssolaires.blog.ca).

The possibility of generating high working temperatures (up to 4000 K) to operate conventional steam engines for electricity production in solar concentrators has been proven (http://www.bine.info/en/topics/energygeneration/publikation/solarthermische-kraftwerke-2/en-passant-aus-der-praxis-i-iii). Moreover, flat-plate collectors are used to generate low-temperature heat (<365 K), which is efficient for producing hot water or heating spaces (http://outilssolaires.blog.ca). However, the biggest disadvantages of the low-temperature heat collectors are the inability to transport the energy for over long distances and the low efficiency if used to produce electricity (Chaabene, 2009).

1.2.2.2 Geothermal energy

Geothermal energy consists in extracting the soil energy on the basis of the temperature increase from the surface to the center of the earth (Fridleifsson, 2001), where the heat is produced by the natural radioactivity of the rocks. The geothermal energy, used to produce electricity, operates in very hot or very deep wells, geothermal sources, where water is injected under pressure into the rock.

Compared to other renewable energies, geothermal energy has the advantage of not depending on atmospheric conditions. It is therefore reliable and available over time (Fridleifsson, 2003; Siegfried, 2014). However, the energy extraction requires a high investment and sophisticated equipment (Siegfried, 2014).

1.2.3 Wind energy

Wind energy is a renewable energy obtained from the pressure difference of natural warm and cool areas, which creates air masses in constant movement (Cavallo, 2007). The electricity from wind is generated by a turbine that converts a portion of the kinetic energy from the wind into a mechanical energy available on a generator shaft (GWEC, 2013).

World wind energy resources are substantial, and in many areas, such as the United States and Northern Europe, could in theory supply all of the electricity demand. However, the remote or challenging locations, the intermittent character of the wind resources, and the necessity of long distances for energy transmission are considered the main drawbacks of wind energy (Pelc & Fujita, 2002).

1.2.4 Wave energy

Produced by wind action, wave energy is considered as an indirect form of solar energy (Clément et al., 2002). In fact, wind generates waves. When arriving at wave energy converters, these waves cede some of their energy that is converted into electricity. Similarly to wind energy, the main drawback of wave energy is its variability on several timescales (Chanson, 1994): from wave to wave, with the state of the sea, and from month to month. Hence, unfortunately, the energy recovery is still not profitable.

1.2.5 Hydraulic energy

Electricity is also produced from water flows, especially in dams constructed across rivers (Demirbas, 2005). Since huge water volumes can be stored, dams produce important amounts of clean energy. The high stability of the source and the possibility of using small dams (<10 kW for isolated mini-grid sites) are considered a great advantage of hydraulic energy (Chaabene, 2009). However, the impossibility of constructing many dams is the most important disadvantage of this type of energy.

1.2.6 Biomass energy

Biomass is one of the earliest sources of energy, with very specific properties. Biomass material (vegetable or animal) is transferred into electricity for example by burning waste in specific boilers (Saidur, Abdulaziz, Demirbas, Hossain, & Mekhilef, 2011). Biomass is divided into three categories: dry biomass (wood, agricultural waste, etc.), biogas, and biomass wet (bioethanol, biodiesel, vegetable oil, etc.).

It has been found that using biomass in boilers offers many economic, social, and environmental benefits, such as financial net savings, conservation of fossil fuel resources, and the reduction of carbon dioxide emissions. However, it requires a great harvesting and collection of material. Moreover, the transportation and storage costs are important (Naika, van Lidt de Jeude, de Goffau, Hilmi, & van Dam, 2005).

1.3 RENEWABLE ENERGIES FOR IRRIGATION

The need to save water and energy is a serious issue that has increased in importance over the last years and will become more important in the near future (Henrik, 2007). The low price of fuel was the reason why renewable energy sources are not widely used in several applications, including water pumping. So, pumping systems based on renewable energies are still scarce, even though they have clear advantages, namely, low generating costs, suitability for remote areas, and being environmentally friendly. Nowadays, the price of electric energy is rising constantly, which makes investing in more efficient solutions increasing (Ibrahim, 2000).

Renewable energies have been used in water pump applications, especially in remote agricultural areas, thanks to the potential of renewable energies. The renewable energies use depends on the user propensity to invest in renewable based pumping systems, his/her awareness and knowledge of the technology for water pumping, and also on the availability, reliability, and economics of conventional options (Johansson et al., 1992). Moreover, the evaluation of the groundwater volume required for irrigation and its availability in the area are also relevant in determining the profitability of renewable energies. Photovoltaic powered electric water pumping systems (PPEWPS) and wind powered electric water pumping systems (WPEWPS) are the most common installations used for water pumping (Johansson et al., 1992). PPEWPS are promising solutions, especially in small-scale installations in regions characterized by good amounts of solar energy over the year (Kumar & Kandpal, 2007; Purohit & Kandpal, 2005). In fact, it is recommended that, for installing solar photovoltaic pumps, the average daily solar radiation in the least sunny month should be greater than 3.5 kW/m^2 on a horizontal surface (Henrik, 2007). Thanks to their efficiency and cost-efficiency rate, PPEWPS have been very popular and they have been developed to appear in these following categories (Purohit & Kandpal, 2005):

- *Directly coupled PPEWPS*: These systems pump water only when the photovoltaic modules capture the solar radiation.
- *Maximum Power Point PPEWPS*: These installations include MPP trackers to enhance the panels efficiency and thus increase the pumped water volume.

- *Batteries PPEWPS*: These systems include batteries to supply pumps when the panels power generation is not sufficient.
- *Sun trackers PPEWPS*: These installations include sun trackers to maximize the solar energy received. They are considered expensive and complicated (Johansson et al., 1992).

WPEWPS have been used in windy sites and can be classified as:

- *DC type WPEWPS*: This category of WPEWPS produces AC energy via wind turbines, which is then rectified to DC and used to supply DC loads (Purohit & Kandpal, 2005).
- *AC type WPEWPS*: These installations generate AC energy used directly to supply AC loads (IEA-AIE, 2013; Sumaila et al., 2008). Moreover, they can take the form of a DC type when they are small size WPEWPS. For instance, permanent magnet synchronous generators with embedded rectifiers are used in small size and fixed pitch wind turbines, which have a simpler construction and are less expensive than DC type WPEWPS.

Some installations combine solar panels and wind turbines to compensate the solar radiation and the wind velocity fluctuations. These sources act in a complementary way, since, generally, when the solar radiation is high, the wind velocity is low. This combination may result in a more reliable but complex water pumping, since electric power generated by wind turbines is highly erratic and may affect both the power quality and the planning of power systems (Ben Ammar, 2011).

Hence, as has been shown, there is a multitude of systems based on renewable energies. However, the choice of the energy source for the pump supply depends essentially on the site characteristics and the water needed by the crops. For the target application, photovoltaic system with MPPT and batteries will be selected. The irrigation methods are now detailed.

1.4 TOMATOES IRRIGATION

Generally, drip and furrow irrigation are the most used methods for tomatoes irrigation (Nema et al., 2009). Although mulching irrigation contributes to crop production by way of influencing soil productivity and weed control (Bernal-Agustín & Dufo-López, 2009; Deshmukh & Deshmukh, 2008), drip irrigation, characterized by its suitability for small and frequent irrigation applications (Nema et al., 2009), is selected here, since it only requires a small water volume and it allows the fruit production to be increased (Chaabene, 2009). Small but frequent water applications enable the plant to grow well, without any effect from water stress, thanks to the frequent water applications between consecutive irrigation periods (Nema et al., 2009; Yahyaoui et al., 2016).

Indeed, several researchers have focused on the yield improvement by drip irrigation of various crops (especially tomato). In fact, it has been reported that drip irrigation allows 30–50% higher tomato yields (Nema et al., 2009) and its use, either alone (or in combination with mulching methods), increases the tomato yield over the normal method of irrigation, which represents 44% savings in irrigation water (Nema et al., 2009). Thus, the irrigation method generally affects the yield production.

1.4.1 Climatic study

Tomato is not resistant to drought. Hence, its yield decreases considerably after short periods of water deficiency. The regularity in watering the plants is important, especially during flowering and fruit formation (Ismail, Ozawa, & Khondaker, 2007). The needed water amount depends on the type of the soil and on the weather (amount of rain, humidity, and temperature) (Ismail et al., 2007). In Tunisia, generally farmers use furrow or drip irrigation, which is a common method for irrigating tomatoes, thanks to its economic advantages in saving water and increasing the yield production (Raes, Sahli, Van Looij, Ben Mechlia, & Persoons, 2000). Farmers adopt this technique for greenhouses and for outdoor cultivation for which, the frequency of tomatoes irrigation depends on the growing stage of the plant and the rainfall. In this sense, Allen et al. (2006) established a guideline for the irrigation frequency and duration, to adjust correctly the irrigation pattern to the actual weather and the water-limiting conditions in Tunisia. This study allows:

- adjusting the irrigation frequency to the actual weather conditions throughout the growing season,
- selecting the irrigation duration as a function of the irrigation installation characteristics.

The irrigation calendars establishment requires a good knowledge of the meteorological parameters of the target region. Among them, the reference crop evapotranspiration (ET_0) and the rainfall levels, which can be expected for a given 10-day period (Allen et al., 2006; Yahyaoui et al., 2016). Since the irrigation objective chart is to adjust the irrigation calendar to the actual weather conditions, several irrigation calendars are developed, using various probability levels for rainfall and ET_0. Hence, four weather conditions can be distinguished (Allen et al., 2006):

1. hot weather conditions without any rainfall (20% ET_0 and no rain),
2. dry (40% ET_0 and 80% rain),
3. normal (mean ET_0 and rain),
4. humid (60% ET_0 and 20% rain).

1.4.2 Soil data

Medjez El Beb, which is the target area of the PV water pumping installation, has a clay loam soil type, which is characterized by a water volume content at field capacity and wilting point (which is the water level below the plant is shriveled) equal to, respectively, 42% and 26%. The corresponding total available soil water is 160 mm (water)/m (soil depth) and the infiltration rate is 100 mm/day.

1.4.3 Crop data

Tomatoes harvested in Tunisia during the summer period are sown in nursery plants during *February*. In *March*, the seedlings are transplanted in the fields. Eight to ten weeks after sowing, flowering occurs in the middle of *May*. At the end of this month and at the beginning of *June*, fruits ripening occurs. In *July*, the fruits are ready to be harvested (Allen et al., 2006).

1.4.4 Irrigation intervals

The net irrigation requirement is obtained by subtracting from the crop water requirement (ET_0) the expected rainfall volume. Hence, the irrigation interval can be derived from the calculated irrigation requirement by means of soil characteristics. Generally, the irrigation is frequent during peak periods when the crop water demand is high and rainfall is small. Thus, the irrigation is less frequent when ET_0 is small or rainfall is frequent (Allen et al., 2006).

The chart presents guidelines to:

- adjust the irrigation interval to the varying climatic conditions during the growing season,
- select the irrigation duration as a function of type, layout, and efficiency of the drip system.

The guidelines are based on information concerning the actual weather conditions, local and technical aspects of the irrigation system, and the crop response to water. The combination of all this information results in an irrigation calendar that is specific for a given farm and adjustable to the actual weather conditions (Tables 1.1 and 1.2).

Some guidelines were proposed to adjust the irrigation frequency to the actual weather conditions throughout the growing season (Allen et al., 2006). Since little or no rainfall is expected during the summer period, the farmer is advised to irrigate daily in *June*. During the ripening stage (*July*), the crop is less sensitive to water stress and the irrigation interval may be increased to 2 days. At the middle of *April*, during the crop development stage, the irrigation interval depends on the actual weather conditions: when it is hot and it does not rain or it is rather dry, it is recommended to irrigate every 2 days. Under normal rainfall conditions, the irrigation interval might be increased to 3 days. In practice, as *April* is wet and the rains are well

Table 1.1 Irrigation chart for drip irrigated tomatoes in the region of Tunis

Month		**March**			**April**			**May**			**June**			**July**		
Decade		**1**	**2**	**3**	**1**	**2**	**3**	**1**	**2**	**3**	**1**	**2**	**3**	**1**	**2**	**3**
Meteorological conditions	Hot + Dry	3		2			1								2	
	Dry	4		3		2		1							2	
	Normal				3		2		1						2	
	Wet						3	2			1				2	
Growth period		Establishment			Vegetative			Flowering			Yield formation			Ripening		
Sensibility to water stress		Sensible			Moderate			Very sensible			Sensible			Moderate		

Table 1.2 Irrigation duration

Discharge (L/h · m²)	**Irrigation efficiency**	
	Good (90%)	**Medium (70%)**
2	3 h 30 min	4 h 15 min
3	2 h 15 min	3 h 00 min
4	1 h 45 min	2 h 15 min
5	1 h 30 min	1 h 45 min
6	1 h 15 min	1 h 30 min
7	1 h 00 min	1 h 15 min
8	1 h 00 min	1 h 00 min
9	0 h 45 min	1 h 00 min
10	0 h 45 min	1 h 00 min

distributed, no irrigation is required during most of the month since rainfall provides the crop water requirements.

Many studies show that the crop is most sensitive to water deficit during and immediately after transplanting, during flowering, and yield formation (Allen et al., 2006). Moreover, moderate water deficit during the vegetative period enhances the root growth. However, water deficit during the flowering period causes the flowers drop.

Given the type of the installation, the distance between emitters on the lateral, the lateral spacing, and the emitter discharge, the discharge per unit area can easily be calculated (Fig. 1.3). Consequently, using this method, the total crops consumption of water for every 10 days can be calculated.

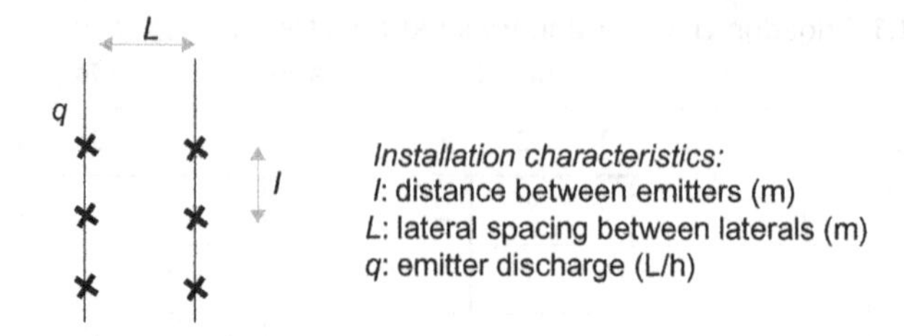

Figure 1.3 Calculation of the discharge per unit surface ($L/h \cdot m^2$).

1.5 CONCLUSIONS

A study of the state of the art of renewable energies used for irrigation has been discussed. The situation of using renewable energies worldwide, namely, photovoltaic, thermal, wind, wave, hydraulic, and biomass energies, and the different photovoltaic installation architectures that can be used have been studied in depth.

Knowing the specification of the studied site, it is obvious to use a photovoltaic based installation to irrigate a plot of land planted with tomatoes. The installation efficiency depends essentially on the system sizing and the energy management. But before all, the system components models should be studied in depth, which will be described in Chapter 2, Modeling of the Photovoltaic Irrigation Plant Components.

REFERENCES

Allen, R. G., Pruitt, W. O., Wright, J. L., Howell, T. A., Ventura, F., Snyder, R., et al. (2006). A recommendation on standardized surface resistance for hourly calculation of reference ET by the FAO56 Penman-Monteith method. *Agricultural Water Management*, *81*(1), 1–22.

Belakhal, S. (2010). Conception & Commande des Machines à Aimants Permanents Dédiées aux Energies Renouvelables (thesis). Algeria: University of Constantine.

Ben Ammar, M. (2011). Contribution à l'optimisation de la gestion des systèmes multi-sources d'énergies renouvelables (thesis). Tunisia: National Engineering School of Sfax.

Bernal-Agustín, J. L., & Dufo-López, R. (2009). Simulation and optimization of stand-alone hybrid renewable energy systems. *Renewable and Sustainable Energy Reviews*, *13*(8), 2111–2118.

Cavallo, A. (2007). Controllable and affordable utility-scale electricity from intermittent wind resources and compressed air energy storage (CAES). *Energy*, *32*(2), 120–127.

Chaabene, M. (2009). Gestion énergétique des systèmes photovoltaïques (Master course). Tunisia: National School for Engineers of Sfax.

Chanson, H. (1994). Comparison of energy dissipation between nape and skimming flow regimes on stepped chutes. *Journal of Hydraulic Research*, *32*(2), 213–218.

Chapman, N. A., & Mc Kinley, I. G. (1987). *The geological disposal of nuclear waste*. New York, NY: John Wiley & Sons, Inc., ISBN 0-471-91249-2.

Chenni, R., Messaoud Makhlouf, M., Kerbache, T., & Bouzid, A. (2007). A detailed modelling method for photovoltaic cells. *Energy*, *32*(9), 1724–1730.

Clément, A., McCullen, P., Falca, A., Fiorentino, A., Gardner, F., Hammarlund, K., et al. (2002). Wave energy in Europe: current status and perspectives. *Renewable and Sustainable Energy Reviews*, *6*(5), 405–431.

de Brito, M. A. G., Galotto, L., Sampaio, L. P., de Azevedo Melo, G., & Canesin, C. A. (2013). Evaluation of the main MPPT techniques for photovoltaic applications. *IEEE Transactions on Industrial Electronics*, *60*(3), 1156–1167.

Demirbas, A. (2005). Potential applications of renewable energy sources, biomass combustion problems in boiler power systems and combustion related environmental issues. *Progress in Energy and Combustion Science*, *31*(2), 171–192.

Deshmukh, M. K., & Deshmukh, S. S. (2008). Modeling of hybrid renewable energy systems. *Renewable and Sustainable Energy Reviews*, *12*(1), 235–249.

EPIA (2014). *Global market outlook for photovoltaic 2014–2018*. European Photovoltaic Industry Association. <www.epia.org>.

Fridleifsson, I. B. (2001). Geothermal energy for the benefit of the people. *Renewable and Sustainable Energy Reviews*, *5*(3), 299–312.

Fridleifsson, I. B. (2003). Status of geothermal energy amongst the world's energy sources. *Geothermics*, *32*(4), 379–388.

Goswami, D. Y., Kreith, F., & Kreider, J. F. (2000). *Principles of solar engineering*. CRC Press, ISBN 1-56032-714-6.

Green, M. A. (1982). *Solar cells: operating principles, technology, and system applications*. Englewood Cliffs, NJ: Prentice-Hall, Inc.

Gules, R., De Pellegrin Pacheco, J., Hey, H. L., & Imhoff, J. (2008). A maximum power point tracking system with parallel connection for PV stand-alone applications. *IEEE Transactions on Industrial Electronics*, *55*(7), 2674–2683.

GWEC (2013). *Global wind statistics*. Global Wind Energy Council.

Henrik, L. (2007). Renewable energy strategies for sustainable development. *Energy*, *32*, 912–919.

Ibrahim, D. (2000). Renewable energy and sustainable development: a crucial review. *Renewable and Sustainable Energy Reviews*, *4*, 157–175.

IEA-AIE (2013). *Key world energy statistics 2013*. International Energy Agency. www.iea.org.

SM Ismail; K Ozawa; & NA Khondaker. (2007). Effect of irrigation frequency and timing on tomato yield, soil water dynamics and water use efficiency under drip irrigation. *Proceedings of the eleventh international water technology conference* (pp. 15–18).

Johansson, T. B., Kelly, H., Amulya, R. K. N., & Williams, R. H. (1992). *Renewable energy: sources for fuels and electricity*. Island Press, ISBN 1-55963-139-2.

Kumar, A., & Kandpal, T. C. (2007). Renewable energy technologies for irrigation water pumping in India: a preliminary attempt towards potential estimation. *Energy*, *32*(5), 861–870.

Lutze, W., & Ewing, R. C. (1988). *Radioactive waste forms for the future*. Amsterdam, Netherlands: North Holland, ISBN 0 444 87104 7.

Naika, S., van Lidt de Jeude, J., de Goffau, M., Hilmi, M., & van Dam, B. (2005). *Cultivation of tomato*. Wageningen: Agromisa Foundation.

Nema, P., Nema, R. K., & Rangnekar, S. (2009). A current and future state of art development of hybrid energy system using wind and PV-solar: a review. *Renewable and Sustainable Energy Reviews*, *13*(8), 2096–2103.

Pelc, R., & Fujita, R. M. (2002). Renewable energy from the ocean. *Marine Policy*, *26*(6), 471–479.

Price, S. C., Stuart, A. C., Yang, L., Zhou, H., & You, W. (2011). Fluorine substituted conjugated polymer of medium band gap yields 7% efficiency in polymer – fullerene solar cells. *Journal of the American Chemical Society*, *133*(12), 4625–4631.

Purohit, P., & Kandpal, T. C. (2005). Renewable energy technologies for irrigation water pumping in India: projected levels of dissemination, energy delivery, and investment requirements using available diffusion models. *Renewable and Sustainable Energy Reviews*, *9*(6), 592–607.

Raes, D., Sahli, A., Van Looij, J., Ben Mechlia, N., & Persoons, E. (2000). Chart for guiding irrigation in real time. *Irrigation and Drainage Systems*, *14*, 343–352.

Saidur, R., Abdulaziz, E. A., Demirbas, A. H., Hossain, M. S., & Mekhilef, S. (2011). A review on biomass as a fuel for boilers. *Renewable and Sustainable Energy Reviews*, *15*(5), 2262–2289.

Siegfried, H. (2014). *Grid integration of wind energy.* John Wiley & Sons.

Sumaila, U. R., Teh, L., Watson, R., Tyedmers, P., & Pauly, D. (2008). Fuel price increase, subsidies, overcapacity, and resource sustainability. *ICES Journal of Marine Science, 65*(6), 832–840.

Wang, H. & Zhang, D. (2010). The stand-alone PV generation system with parallel battery charger. *Proceedings of the IEEE conference on electrical and control engineering (ICECE)* (pp. 4450–4453).

W Xiao & Dunford W.G. (2004). A modified adaptive hill climbing MPPT method for photovoltaic power systems. *Proceedings of the IEEE conference on power electronics specialists* (pp. 1957–1963).

Yahyaoui, I., Tadeo, F., & Segatto, M. E. V. (2016). Energy and water management for drip-irrigation of tomatoes in a semi-arid district. *Agricultural Water Management.*

Yotaka, H., Kazuo, Y., & Hitomi, S. (1984). Spectral sensitization in an organic P-N junction photovoltaic cell. *Applied Physics Letters, 45*(10), 1144–1145.

CHAPTER 2

Modeling of the Photovoltaic Irrigation Plant Components

2.1 INTRODUCTION

Following the discussion in the previous chapter, using autonomous photovoltaic (PV) installations is a promising solution to supply water pumping systems for crop irrigation in remote agriculture areas, since the studied isolated site is characterized by a good solar insolation throughout the year (Balghouthi, Chahbani, & Guizani, 2012; Bouadila, Lazaar, Skouri, Kooli, & Farhat, 2014; Soussi, Balghouthi, & Guizani, 2013). After presenting the adopted installation (Section 2.2), the system components models are detailed in Section 2.3 and validated in Section 2.4. These models will be used to define the optimum components size in Chapter 3, Sizing Optimization of the Photovoltaic Irrigation Plant Components and to perform the Energy Management Algorithm in Chapter 4, Optimum Energy Management of the Photovoltaic Irrigation Installation.

Now, the book' author presents for the reader an explanation about the target system characteristics.

2.2 TARGET SYSTEM

As Tunisia climate is considered semiarid (Rana, Katerji, Lazzara, & Ferrara, 2012) and many crops need to be irrigated regularly (Bernstein & Francois, 1973), the use of an autonomous installation for water pumping is required. The characteristics of the installation selected (presented in Fig. 2.1) are explained as follows.

- *Choice of the renewable energy*: Since the land is characterized by a good amount of solar energy during the year (Rana et al., 2012), a PPEWPS installation that includes the MPPT technique and batteries is chosen.
- *Choice of the components*: The installation is composed of photovoltaic panels (PVPs). Since the installation autonomy is required, a lead–acid battery bank is used, as it is efficient and economic. These components supply a centrifuge water pump (Appendix B) driven by an induction machine (IM) (Appendix C), as the application adopted in this book is characterized by a constant flux and a moderate head (Fig. 2.1). The regulator is composed of three relays that allow the components to be connected and disconnected. For the reservoir, its volume is just considered, which is the maximum volume needed by the crops in the most critical month (*July*).

Specifications of Photovoltaic Pumping Systems in Agriculture
DOI: http://dx.doi.org/10.1016/B978-0-12-812039-2.00002-8

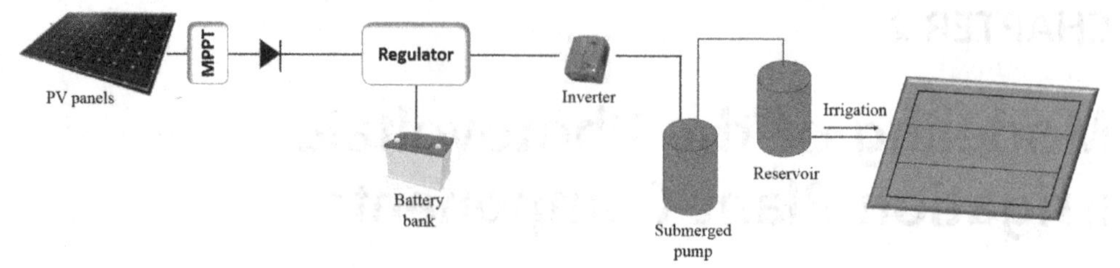

Figure 2.1 Proposed PV irrigation system.

- *Choice of the architecture*: Among the objectives of this book are to determine the optimum size of the system components and control the installation. Hence, a specific parallel configuration for these components has been chosen. The installation cabling is done by DC bus.
- *MPPT technique*: The installation is equipped with MPPT bloc which trucks the Point of the Maximum Power generation (MPP) (Salas, Olias, Barrado, & Lazaro, 2006).
- *Choice of the irrigation method*: As it has been previously discussed in Section 1.4, Tomatoes Irrigation, the drip irrigation is chosen for irrigating tomatoes, thanks to its advantages in enhancing the production yield and saving water and money (Bernstein & Francois, 1973; Sweeeney, Graett, Bottcher, Lacario, & Camphll, 1987).

After choosing the system characteristics, an adequate modeling of the installation components is relevant to the optimum sizing of the system components and the best use of the energy generated, which is detailed in the following section.

2.3 SYSTEM MODELING AND VALIDATION

In order to evaluate the system elements size and optimize the energy generated, an essential step consists in modeling the installation components. Hence, some models for the PVPs, the batteries, and the pump are presented here. Some of which will be experimentally validated and then used for sizing (see chapter: Sizing Optimization of the Photovoltaic Irrigation Plant Components) and management (see chapter: Optimum Energy Management of the Photovoltaic Irrigation Installation).

2.3.1 System modeling

In this section, the installation components models are explained, except the regulator (which is detailed in chapter: Optimum Energy Management of the Photovoltaic Irrigation Installation).

2.3.1.1 PVPs models

In autonomous PV installations, panels are the source that generates the electric energy for the rest of the components. To better understand the PV panel behavior, an essential step consists in studying the parameters affecting the PV power generation. These parameters are essentially the solar radiation G, the ambient temperature T_a, and the panel characteristics (Bernal-Agustín & Dufo-López, 2009; Khatib, Mohamed, & Sopian, 2012; Yahyaoui, Chaabene, & Tadeo, 2013), which are detailed below.

Solar radiation model

Solar radiation data provide information on how much of the sun energy strikes a surface at a location on the Earth during a time period. These data are needed for effective research into solar energy utilization. Solar energy consists of two parts: (1) extraterrestrial solar energy, which is above the atmosphere and (2) global solar energy, which is under the atmosphere (Posadillo & López Luque, 2009). The global solar energy incidence on a tilted panel is generally evaluated using the Liu and Jordan relations (El-Sebaii, Al-Hazmi, Al-Ghamdi, & Yaghmour, 2010; Posadillo & López Luque, 2009). In this model, the solar radiation depends essentially on the position of the sun, which is determined by using the declination and the hour angle of the sun (Bernal-Agustín & Dufo-López, 2009) detailed below.

Solar declination The sun declination δ, needed to determine its position, is the angle between the sun direction at the solar noon and its projection on the equatorial plane. In fact, it reaches its maximum (23.45°) at the summer solstice (21 *June*), and its minimum (−23.45°) at the winter solstice (*December* 21). It is described by Coopers equation (Arunkumar et al., 2012; Bernal-Agustín & Dufo-López, 2009; Moghadam, Tabrizi, & Sharak, 2011):

$$\delta = 23.45 \sin\left(2\pi \frac{284 + d}{365}\right) \tag{2.1}$$

where d is the day number in the year.

Hour angle of the sun The hour angle of the sun w is the sun East to West angular displacement around the polar axis. The value of the hour angle is zero at noon, negative in the morning, and positive in the afternoon and it is increased by 15° per hour.

The hour angle of the sun w_s at sunset is given by (Şen, 2008):

$$\cos w_s = -tg\varphi tg\delta \tag{2.2}$$

where δ is the declination (°) calculated from Eq. (2.1) and φ is the site latitude (°).

Table 2.1 Numerical parameters for TE500CR and Sunel panels

Parameters	Values (TE500CR)	Values (Sunel)
V_c	22.3 V	36.7 V
$I_{scT_{ref}}$	4.2 A	8.6 A
n_s	36 cells/module (Ben Ammar, 2011)	60 cells/module
A	0.095%/K (Ben Ammar, 2011)	0.039%/K
V_g	1.12 V (Chaabene, 2009)	
η_r	13% (Chaabene, 2009)	
β_{pv}	0.4% (Chaabene, 2009)	
T_{ref}	25°C (Chaabene, 2009)	
NOCT	45°C (Chaabene, 2009)	
K_B	1.3806×10^{-23} J/K (Chaabene, 2009)	
Q	1.6×10^{-16} C (Chaabene, 2009)	
G_{sc}	1367 W/m^2 (Chaabene, 2009)	
G_{ref}	1000 W/m^2 (Chaabene, 2009)	

Extraterrestrial radiation and clearness index The extraterrestrial solar radiation H_0 is the solar radiation outside the Earth atmosphere. The extraterrestrial radiation $H_0(d)$ (J/m^2) on a horizontal surface for the day d is obtained using the following equation (Clean Energy Decision Support Centre, 2001–2004):

$$H_0(d) = \frac{24 \times 3600}{\pi} G_{sc}\left(1 + 0.033\cos\left(2\pi\frac{d}{365}\right)\right)(\cos\varphi\cos\delta + w_s\sin\varphi\sin\delta) \tag{2.3}$$

where G_{sc} is the solar constant (Table 2.1).

The solar radiation is attenuated by the atmospheric layer and clouds before it reaches the Earth soil. Hence, the quotient between the ground and the extraterrestrial solar radiations, known as clearness index K_t can be obtained. The monthly average of this index is defined by (Clean Energy Decision Support Centre, 2001–2004; Orgill & Hollands, 1977):

$$\overline{K_t} = \frac{\overline{H}}{\overline{H_0}} \tag{2.4}$$

where:

$\overline{H}$: the monthly average of the solar radiation on a horizontal plane,

$\overline{H_0}$: the monthly average of the extraterrestrial radiation on a horizontal plane.

Solar radiation calculation Solar radiations do not strike the Earth surfaces perpendicularly during all the year, due to the Earth declination (Şen, 2008). Hence, it is obvious to incline the solar panels, in order to maximize the amount of solar radiations that reach the PVP perpendicularly. The total solar radiation on a tilted PVP is calculated as follows:

1. Calculation of the diffused, the global and the direct solar radiation in a horizontal panel following Eqs. (2.5–2.13).
2. Calculation of the global solar radiation corresponding to a tilted panel following Eqs. (2.14–2.17).
3. Sum of the hourly values of the solar radiation following Eq. (2.18).

These two steps are presented in detail:

1. *Calculation of the diffused and global solar radiation*

In the literature, several models have been used to determine the diffused solar radiation $\overline{H_d}$. For instance, El-Sebaii et al. (2010) used models that classify the daily diffused solar radiation based on the daily clearness index intervals. Other works proposed a seasonal relation for $\overline{H_d}$ that depends on the sun hour angle at sunset at the month mean day (Chaabene, 2009; Erbs, Klein, & Duffie, 1982). The monthly diffused solar radiation has also been developed using the monthly clearness index (Duffie & Beckman, 2013).

Hence, since a monthly radiation average is needed for the sizing of the installation components, the diffused and global solar radiations are deduced using the monthly average global solar radiation in a horizontal panel. The diffused insolation $\overline{H_d}$ is a function of the hour angle at sunset. It is described as follows (Clean Energy Decision Support Centre, 2001–2004; Duffie & Beckman, 2013; Kenny, Friesen, Chianese, Bernasconi, & Dunlop, 2003; Orgill & Hollands, 1977):

- If w_s is less than 81.4°:

$$\overline{H_d} = \overline{H}\left(1.391 - 3.56\,\overline{K_t} + 4.189\,\overline{K_t}^2 - 2.137\,\overline{K_t}^3\right) \tag{2.5}$$

- If w_s is higher than 81.4°:

$$\overline{H_d} = \overline{H}\left(1.311 - 3.022\,\overline{K_t} + 3.427\,\overline{K_t}^2 - 1.821\,\overline{K_t}^3\right) \tag{2.6}$$

The hourly diffused and global insolation H_d and H are respectively obtained using Eqs. (2.7) and (2.8) (Duffie & Beckman, 2013):

$$H_d(t, d) = r_d(t, d)\overline{H_d} \tag{2.7}$$

$$H(t, d) = r_t(t, d)\overline{H} \tag{2.8}$$

where r_d is the ratio of the hourly to daily total diffuse solar radiation expressed by (Clean Energy Decision Support Centre, 2001–2004; Duffie & Beckman, 2013; Orgill & Hollands, 1977):

$$r_d(t, d) = \frac{\pi}{24}\frac{\cos w - \cos w_s}{\sin w_s - w_s}\cos w_s \tag{2.9}$$

where:

w: the hour angle of the sun,

w_s: the hour angle of the sun at sunset (following Eq. (2.8)),

$r_t(t, d)$: the ratio of the hourly to the daily total global solar radiation, expressed by (Clean Energy Decision Support Centre, 2001–2004; Duffie & Beckman, 2013; Orgill & Hollands, 1977):

$$r_t(t,d) = \frac{\pi}{24} \frac{\cos w - \cos w_s}{\sin w_s - w_s \cos w_s} (a + b \cos w) \tag{2.10}$$

where:

$$a = 0.409 + 0.501 \sin\left(w_s - \frac{\pi}{3}\right) \tag{2.11}$$

$$b = 0.6609 + 0.4767 \cos\left(w_s - \frac{\pi}{3}\right) \tag{2.12}$$

Hence, the direct solar radiation $H_b(t,d)$ is obtained using the following equation (Duffie & Beckman, 2013; Orgill & Hollands, 1977):

$$H_b(t,d) = H(t,d) - H_d(t,d) \tag{2.13}$$

2. *Calculation of the hourly radiation on a tilted panel*

The total daily solar radiation H_t in a tilted panel is evaluated by varying the hour angle that corresponds to the length of the day. It is expressed by (Chaabene, 2009) (Fig. 2.2):

$$H_t(t,d) = R_b' \, H_b(t,d) + \left(\frac{1 + \cos\beta}{2}\right) H_d(t,d) + \rho \left(\frac{1 - \cos\beta}{2}\right) H(t,d) \tag{2.14}$$

where:

ρ: the albedo of the soil,

β: the panel declination (°),

R_b': the ratio of the direct radiation on the tilted panel and the direct radiation on the horizontal panel, expressed by (Orgill & Hollands, 1977; Şen, 2008):

$$R_b' = \frac{\cos\theta}{\cos\theta_z} \tag{2.15}$$

where:

θ: the radiation incidence angle (°), expressed by:

$$\begin{aligned}\cos\theta = {} & \sin\delta \sin\varphi \cos\beta - \sin\delta \cos\varphi \sin\beta \cos\gamma + \cos\delta \cos\varphi \cos\beta \cos w \\ & + \cos\delta \sin\varphi \sin\beta \cos\gamma \cos w + \cos\delta \sin\beta \sin\gamma \sin w\end{aligned} \tag{2.16}$$

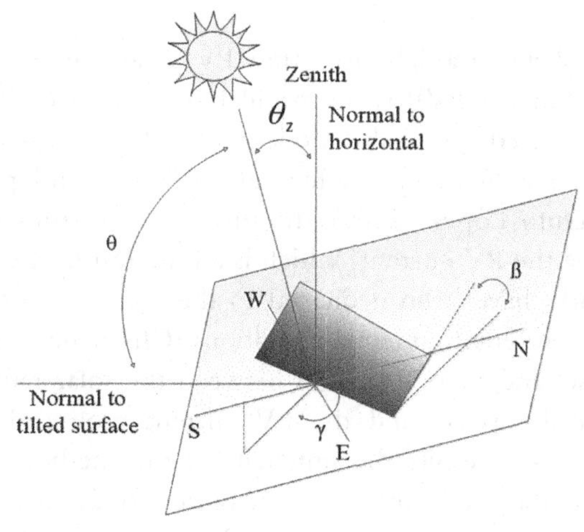

Figure 2.2 Solar radiation angles.

θ_z: the zenith angle of the sun (°), given by:

$$\cos \theta_z = \sin \delta \sin \varphi + \cos \delta \cos \varphi \cos w \tag{2.17}$$

3. *Sum of the hourly values of the solar radiations*

The evaluation of the solar energy W_{pv} is performed by summing the solar radiation received during the day. Hence, it is assumed that during the hour, the solar radiation is constant (Chaabene, 2009). Hence, the solar energy (Wh) is expressed by:

$$W_{pv} = \sum_{t_{sr}}^{t_{ss}} H_t(t)dt \tag{2.18}$$

where:

t_{sr}: the time of sunrise,

t_{ss}: the time of sunset.

Ambient temperature distribution model

The distribution model used to forecast the ambient temperature $T_a(t,d)$ of the day d at the hour t depends on the minimum and the maximum temperatures $T_{min}(d)$, $T_{max}(d)$ of the day d. Thus, $T_a(t,d)$ is expressed by (Chaabene, 2009):

$$T_a(t,d) = \frac{T_{max}(d) + T_{min}(d)}{2} + \frac{T_{max}(d) - T_{min}(d)}{2} \cos\left(\pi \frac{t-13}{24}\right) \tag{2.19}$$

PVPs model

In the literature, several models for the PVPs are used (Chaabene, 2009; Kenny et al., 2003). Since a PVP is the parallel association of PV modules constituted of serially connected PV cells, modeling a PVP consists first in modeling the PV cells and then applying the effect of the series and parallel connections (Bernal-Agustín & Dufo-López, 2009). In this sense, various models have been established to describe the PV current, which is a function of the PV voltage.

In fact, many works have been dedicated to the collection of a number of I–V characteristics in different environmental conditions (Chaabene, 2009). These data are arranged in a database, and then a relation between the solar radiation, the ambient temperature at the cell surface, and the PV current produced is deduced (Xiao, Dunford, & Capel, 2004). Despite the simplicity of this method, it remains practical only for the studied module and cannot be generalized for other modules types.

Hence, researchers use general models, namely, nonlinear models. In this context, some works concentrate on the one (or two) diode based model, which associates a current source, that generates the current I_{ph}, in parallel with one (or two) diode, a parallel resistance called R_p connected in parallel and a serial resistance called R_s, which is connected in series, to describe the current generated by the PV cell (Adamo, Attivissimo, Di Nisio, & Spadavecchia, 2011; González-Longatt, 2005). This model includes parameters of the PV module, namely, the diode quality factor n, the short circuit current and the open circuit voltage $V_{oc\_T_{a\ ref}}$ at the temperature of reference $T_{a\ ref}$. This model allows the PV current I_{pv} in a PV module to be obtained, and therefore, the PV power generated by a PVP composed of parallel PV modules and serial PV cells to be deduced (Fig. 2.3).

Moreover, more generalized models have been developed, such as the Photovoltaic panel (PVP) yield-based model (Chaabene, 2009; Yahyaoui et al., 2013). Indeed, the PVP yield is evaluated using the PV cell parameters values, namely, the temperature coefficient for the panel yield, the panel yield at the reference temperature, etc., and the cell temperature module, which depends on the nominal operating cell temperature (NOCT) and the total solar radiation that strikes the tilted PVP during the day (Chaabene, 2009; Yahyaoui et al., 2013).

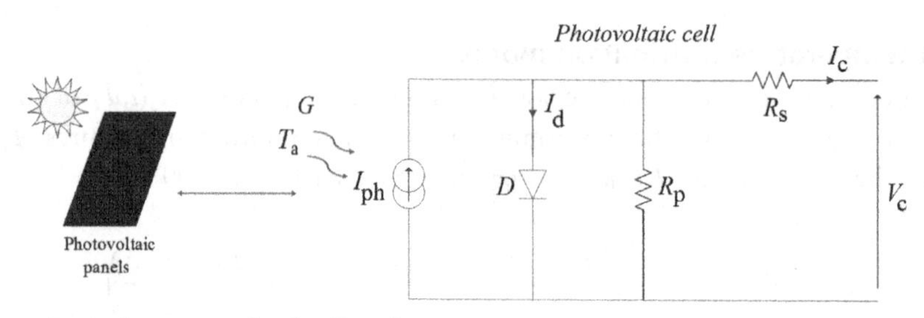

Figure 2.3 Equivalent circuit for the PV cell.

Hence, the yield and the nonlinear models for the PVP are detailed below.

Panel yield based model This simplified model is based on few parameters. The panel yield model is given by Eq. (2.20) (Chaabene, 2009; Yahyaoui et al., 2013):

$$\eta_{pv}(t) = \eta_r(1 - \beta_{pv}(T_c(t) - T_{ref})) \tag{2.20}$$

where:

η_r: the panel yield at the reference temperature,

β_{pv}: the temperature coefficient for the panel yield (°C^{-1}),

$T_c(t)$: the cell temperature (°C),

T_{ref}: the reference temperature (°C).

The cell temperature $T_c(t)$ can be calculated as follows (Chaabene, 2009; Duffie & Beckman, 2013):

$$T_c(t) = T_a(t) + H_t(t, d)\frac{\text{NOCT} - T_{a\ ref}}{800} \tag{2.21}$$

where:

T_a: the ambient temperature (°C),

$H_t(t, d)$: the solar radiation on the tilted panel (W/m^2),

NOCT: the nominal operating cell temperature (°C),

$T_{a\ ref}$: the reference ambient temperature (°C).

Finally, the PV power can be evaluated by Eq. (2.22) (Duffie & Beckman, 2013; Yahyaoui et al., 2013):

$$P_{pv}(t) = S\, H_t(t, d)\eta_{pv}(t) \tag{2.22}$$

where S is the panel surface (m^2).

Panel nonlinear model A one-diode based nonlinear model is used for the management algorithm, using the ideality factor n to describe the diode performance (Chenni, Makhlouf, Kerbache, & Bouzid, 2007; De Blas, Torres, Prieto, & Garcia, 2002). The model uses the solar radiation $G(t)$, the ambient temperature $T_a(t)$ at the panel surface, and the PVP parameters, to evaluate the PV current $I_c(t)$ of one PV cell (see Fig. 2.3) (González-Longatt, 2005). The model is described by Eqs. (2.23)–(2.27) (Chenni et al., 2007; De Blas et al., 2002):

$$I_c(t) = I_{ph}(t) - I_r(t)\left(\exp\left(\frac{V_c(t) + R_s I_c(t)}{V_{t_{T_a}}}\right) - 1\right) - \frac{V_c(t) + R_s I_c(t)}{R_p} \tag{2.23}$$

$$I_{ph}(t) = \frac{G(t)}{G_{ref}} I_{sc}(t) \tag{2.24}$$

$$I_{sc}(t) = I_{sc\_T_{ref}}(1 + a(T_a(t) - T_{a\ ref})) \tag{2.25}$$

$$I_r(t) = I_{r\_T_{a\ ref}} \left(\frac{T_a(t)}{T_{a\ ref}}\right)^{\frac{3}{n}} \exp\left(\frac{-qV_g}{nK_B}\left(\frac{1}{T_a(t)} - \frac{1}{T_{a\ ref}}\right)\right) \tag{2.26}$$

$$I_{r\_T_{a\ ref}} = \frac{I_{sc\_T_{a\ ref}}}{\exp\left(\frac{qV_{c\_T_{a\ ref}}}{nK_B T_{a\ ref}}\right) - 1} \tag{2.27}$$

where:

$I_c(t)$: the estimated PV cell current (A),
$I_{ph}(t)$: the generated photo-current at a given irradiance G (A),
$I_r(t)$: the reverse saturation current for a given temperature T_a (A),
$V_c(t)$: the open circuit voltage of the PV cell (V),
R_s: the serial resistance of the PV module (Ω),
$V_{t\_T_a}$: the thermal potential at the ambient temperature (V),
R_p: the parallel resistance of the PV module (Ω),
G_{ref}: the solar radiation at reference conditions (W/m^2),
$I_{sc}(t)$: the short circuit current for a given temperature T_a (A),
$I_{sc\_T_{a\ ref}}$: the short circuit current per cell at the reference ambient temperature (A),
a: the temperature coefficient for the short circuit current (K^{-1}),
$T_a(t)$: the ambient temperature at the panel surface (K),
$T_{a\ ref}$: the reference temperature at the panel surface (K),
$I_{r\_T_{ref}}$: the reverse saturation current for the reference temperature $T_{a\ ref}$(A),
n: the quality factor,
q: the electron charge (C),
V_g: the gap energy (eV),
K_B: the Boltzmann coefficient (J/K),
$V_{c\_T_{a\ ref}}$: the open circuit voltage per cell at the reference temperature (V).

The PV power P_{pv} generated by the panel is then described as follows (Chaabene, 2009; Yahyaoui et al., 2013):

$$P_{pv}(t) = n_s n_p V_c(t)\left(I_{ph}(t) - I_r(t)\left(\exp\left(\frac{V_c(t) + R_s I_c(t)}{V_{tT_a}}\right) - 1\right) - \frac{V_c(t) + R_s I_c(t)}{R_p}\right) \tag{2.28}$$

where:

n_s: the number of serial PV cells,
n_p: the number of parallel PV modules.

MPPT techniques As is well known, the PV power P_{pv} of the PV system is a nonlinear function crucially influenced by the solar irradiation G and the ambient temperature T_a (Faranda & Leva, 2008). Consequently, the PV system operating point must change to maximize the energy produced. For this, MPPT techniques are used to search for the MPP, at which the PV array must be maintained operating (Salas et al., 2006).

In this sense, researchers have developed many methods for MPPT, such as the Lookup Table (Charfi & Chaabene, 2014), the Neuro-Fuzzy (Veerachary & Yadaiah, 2000), the Incremental Conductance (Veerachary & Yadaiah, 2000), the Perturbation and Observation (P&O) (Femia, Petrone, Spagnuolo, & Vitelli, 2005) methods, and so many others. Generally, they differ in complexity and tracking accuracy, but they all require the sensing of the PV current I_{pv} and voltage V_{pv}, using off-the-shelf hardware. These techniques allow the MPP to be tracked thanks to the use of converters such as choppers, which are controlled by varying their duty cycle α (Charfi & Chaabene, 2014; Yahyaoui et al., 2016) (Fig. 2.4). Some of these methods are now briefly revised.

The lookup table MPPT The lookup table MPPT method consists in dividing the possible solar radiation G and ambient temperature T_a into intervals, then attributing the minimum value of the corresponding interval for the measured climatic data (Charfi & Chaabene, 2014). Hence, for each set of solar radiation and temperature intervals are assigned offline values of the PV voltage V_{mpp}, current I_{mpp}, and power P_{mpp}. Then, a Proportional Integral (PI) type controller adjusts the duty cycle α of the DC–DC converter, to obtain these predetermined values of current and power (Charfi & Chaabene, 2014; Ghaisari, Habibi, & Bakhshai, 2007).

This offline method allows the oscillations around the MPP to be reduced, with a rapid convergence (Charfi & Chaabene, 2014; Yahyaoui et al., 2016), since it compares the duty cycle value, which corresponds to the operation in the MPP under predetermined climatic data, with the one stored in the control system (Charfi & Chaabene, 2014; Ghaisari et al., 2007). However, choosing the minimum value of each interval of G and T_a gives an operating point near but different from the MPP, which causes power loses.

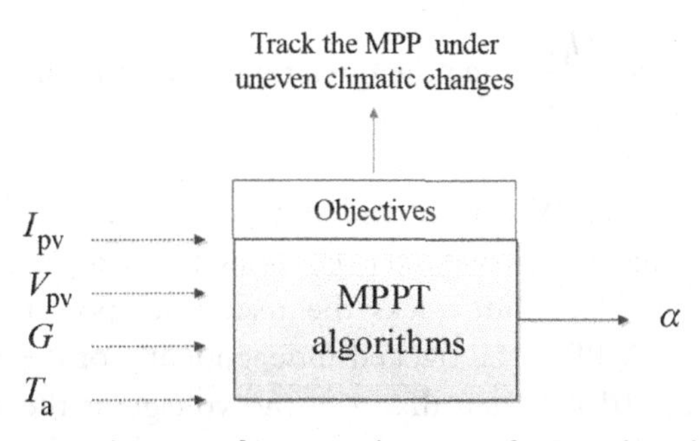

Figure 2.4 General schematic diagram of input and output of MPPT algorithms.

The neuro-fuzzy MPPT The neuro-fuzzy MPPT method is based on training a neuro-fuzzy tool using a solar radiation and ambient temperature database (Charfi & Chaabene, 2014; Salas et al., 2006). Then, the fuzzy rules that describe these relations are deduced. The training step is performed using an artificial neural network (ANN), characterized by the ability to store experimental knowledge, which makes them well suited to tracking the MPP of PVPs (Salas et al., 2006). A multilayer perception network, trained by the back propagation method, is the most widely used technique to calculate the DC–DC optimal duty cycle α, considering the irradiation and the ambient temperature variation (Charfi & Chaabene, 2014).

This method does not require a model for the panel and it can handle nonlinearities. However, it needs a continuous update for the database and a high-performance processor, which makes it a no-economic.

The incremental conductance MPPT The incremental conductance MPPT method uses the current ripple in the chopper output I_{pv} to maximize the panel power P_{pv} using the relation between the current and voltage continuously identified online (Hussein, Muta, Hoshino, & Osakada, 1995; Oi, 2005a). In fact, the incremental conductance for MPPT depends on the array terminal voltage V_{pv}, which is always adjusted according to the desired MPP voltage V_{mpp}, based on the instantaneous and incremental conductance of the PV module. Indeed, the algorithm tests the actual conductance $-\frac{I_{pv}}{V_{pv}}$ and the incremental conductance $\frac{dI_{pv}}{dV_{pv}}$ as follows (Charfi & Chaabene, 2014; Oi, 2005a):

$$\text{If } \frac{dI_{pv}}{dV_{pv}} > -\frac{I_{pv}}{V_{pv}}, \text{ then the operating point is on the left of the MPP} \tag{2.29}$$

so, α is varied to increase V_{pv}.

$$\text{If } \frac{dI_{pv}}{dV_{pv}} < -\frac{I_{pv}}{V_{pv}}, \text{ then the operating point is on the right of the MPP} \tag{2.30}$$

so, α is varied to decrease V_{pv}.

$$\text{If } \frac{dI_{pv}}{dV_{pv}} \approx -\frac{I_{pv}}{V_{pv}}, \text{ then the operating point is in the MPP} \tag{2.31}$$

so, the value of α is maintained.

where dI_{pv} and dV_{pv} are the PV current and voltage variations, respectively.

Hence, by comparing these conductance values following Eqs. (2.29)–(2.31), at each sampling time, the algorithm tracks the maximum power of the PV module. This method allows the MPP to be tracked independently of the module characteristics (Kurella & Suresh, 2013); when $dI_{pv} > 0$, the voltage at the MPP increases and, thus, the algorithm must increase V_{pv} to track V_{mpp}.

Although it has good efficiency, the complexity in implementation remains the main disadvantage of the incremental conductance MPPT method (Salas et al., 2006).

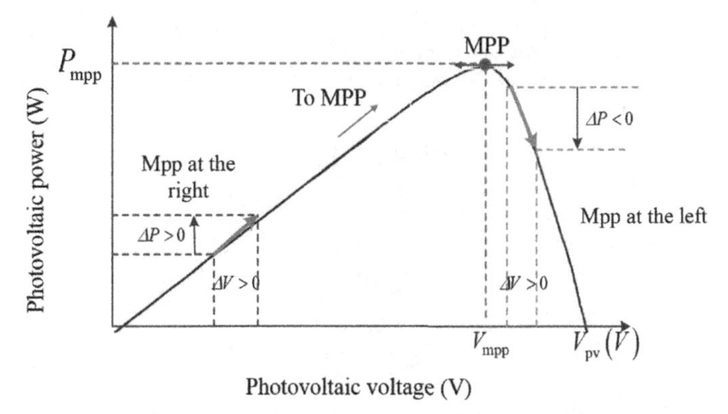

Figure 2.5 Methodology to locate the MPP using the power versus voltage curve.

The P&O MPPT The P&O MPPT method uses the PV current and voltage measurements and compares their previous and present values (Fig. 2.5). In fact, it consists in perturbing the panel voltage and comparing the PV power value obtained with its previous value (Esram & Chapman, 2007). The increase in the PV power generates an increase of the perturbation voltage (Esram & Chapman, 2007). The P&O method is performed as follows (Fig. 2.6) (Desai & Patel, 2007):

$$\text{If } \frac{dP_{pv}}{dV_{pv}} \succ 0, \text{ then the operating point is on the left of the MPP} \tag{2.32}$$

so, α is changed to increase V_{pv}.

$$\text{If } \frac{dI_{pv}}{dV_{pv}} \prec 0, \text{ then the operating point is on the right of the MPP} \tag{2.33}$$

α is changed to decrease V_{pv}.

$$\text{If } \frac{dP_{pv}}{dV_{pv}} \approx 0, \text{ then the operating point is in the MPP} \tag{2.34}$$

so, the value of α is maintained.
where dP_{pv} is the PV power variation, and dV_{pv} is the PV voltage variation.

The oscillations that can be generated by P&O are considered the main drawback of this method (Kurella & Suresh, 2013). However, it is easy to install and its cost is relatively low, so it is the most popular in practice (Kurella & Suresh, 2013).

2.3.1.2 Battery bank

The PVP produces electric energy only when the solar radiation is available. Hence, the use of a battery bank is necessary, to complete the remaining power for the load supply and to store the excess of the PV energy generated.

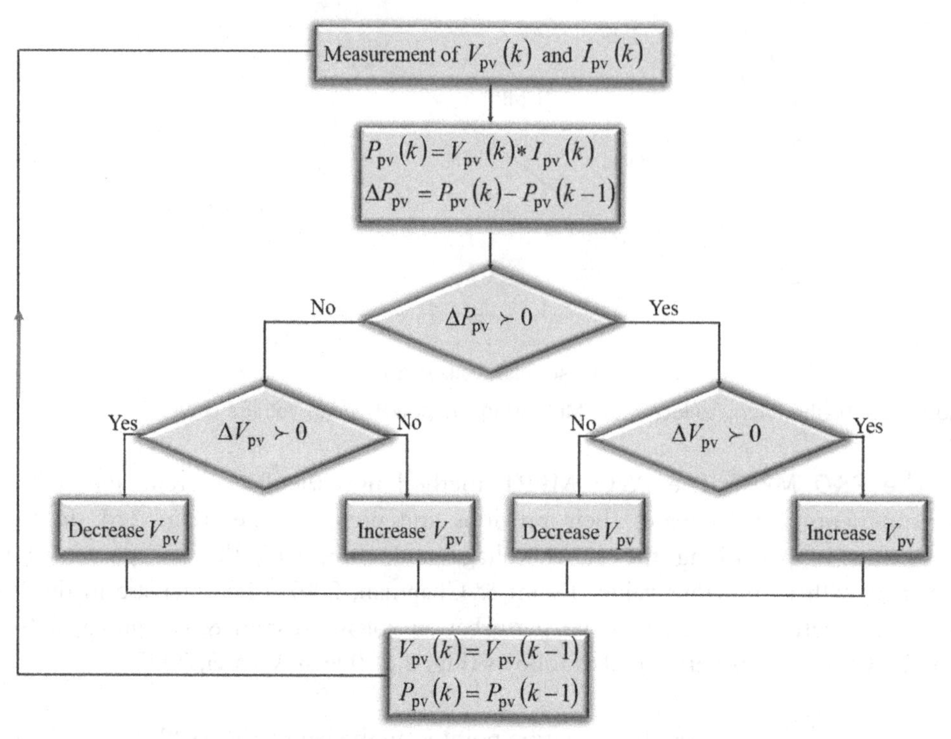

Figure 2.6 Perturb and observe principle for MPPT.

A battery, composed of positive and negative electrodes separated by an electrolyte, converts the chemical energy to electric energy thanks to oxidoreduction reactions (Collins et al., 2010; Zhang et al., 2011). The battery type most used in stationary applications is the lead–acid battery, since the relation between its cost and its life time is acceptable (Collins et al., 2010). Some researchers developed models for the battery characteristics, to describe the battery state of charge (*SOC*) or its depth of discharge (*dod*). In this sense, linear modeling methods, namely, the Coulometric (Chaabene, 2009), the open-circuit voltage methods (Ben Salah & Ouali, 2012; Chaabene, 2009; Fendri & Chaabene, 2012), and the nonlinear dynamic modeling methods have been developed, they are detailed below.

Linear model for the battery behavior description

The most used linear technique for the control of the battery *SOC* consists in measuring and calculating the electricity quantity during the charging and discharging process, and it is called Coulomb-metric method (Chaabene, 2009).

In this method, the battery SOC at an instant k depends directly on the previous SOC at the instant $k-1$, which means that this method requires knowing the initial SOC of the battery and/or when the load is disconnected, in order to know whether the battery was exposed to excessive discharges (Chaabene, 2009). Hence, since the battery open-circuit voltage is proportional to the SOC, it can be also used to describe the SOC. Thus, the linear method for the battery modeling combines the two methods previously detailed: coulomb-metric and open-circuit voltage methods (Chaabene, 2009).

Coulomb-metric method This model considers that the SOC is a result of the electricity remaining in the battery (Ah) by the total battery capacity (Ah). The SOC at the instant $t+1$ is given by Eq. (2.35) (Chaabene, 2009; Yahyaoui et al., 2013):

$$SOC(t+1)=\begin{cases} SOC(t)+\eta_b\dfrac{I_{\mathrm{bi}}(t)}{C_{\mathrm{bat}}} \\ SOC(t)-\dfrac{I_{\mathrm{bo}}(t)}{C_{\mathrm{bat}}} \end{cases} \tag{2.35}$$

where:

η_{b}: mean Faradic yield during a charging cycle,
$I_{\mathrm{bi}}(t)$: the entering battery current (A),
C_{bat}: the maximum battery capacity (Ah),
$I_{\mathrm{bo}}(t)$: the outgoing battery current (A).

The open-circuit method The measuring of the battery open-circuit voltage can be performed in the following cases:

- at the beginning of the battery operating,
- when the SOC is minimum.

In fact, the estimation of the open-circuit voltage must be performed after a reasonable period, necessary for the battery stabilization. It is expressed by Eq. (2.36) (Chaabene, 2009):

$$V_{\mathrm{oc}}=1.64*\left(\frac{V_1-V_0}{0.7}\right)+V_0 \tag{2.36}$$

where:

V_0: the initial open-circuit voltage (V),
V_1: the open-circuit voltage at $t+1$ (V).

Hence, the SOC given by the open-circuit method is as following (Chaabene, 2009):

$$SOC=92.97*V_{\mathrm{oc}}-1064.60 \tag{2.37}$$

Nonlinear model for the battery behavior description

The nonlinear model for modeling the lead–acid battery is characterized by its simplicity: This model has the advantage of using both the battery current and voltage to describe precisely the battery behavior when charging or discharging (Sallem, Chaabene, & Kamoun, 2009a). Its performance is then evaluated from its voltage V_{bat}, its capacity C_p, and its *dod*. In fact, the battery model adopted is composed of a resistance R_t in series with two parallel branches (Ben Salah & Ouali, 2012; Yahyaoui et al., 2013) (Fig. 2.7). The first branch represents the battery storage capacity using a capacitor C_{bulk}, in series with a resistance R_e. The second branch is composed of a capacitor C_s, which represents the diffusion phenomena within the battery, in series with a resistance R_s'. The battery equivalent resistance is described as follows (Ben Salah & Ouali, 2012):

$$R = R_t + \frac{R_e R_s'}{R_e + R_s'} \tag{2.38}$$

where:

R_t is the terminal resistance (Ω), R_e is the end resistance (Ω), and R_s' is the surface resistance (Ω).

The stored charge in the battery C_R is described as follows (Sallem et al., 2009a; Sallem, Chaabene, & Kamoun, 2009b):

$$C_{R_{(k)}} = C_{R_{(k-1)}} + \frac{\partial k}{3600} {I_{bat_{(k)}}}^{k_p} \tag{2.39}$$

where ∂k is the time between instant $k-1$ and k and k_p is the Peukert constant.

The *dod* is given by Eq. (2.40) (Sallem et al., 2009a, 2009b; Yahyaoui et al., 2013):

$$dod_{(k)} = 1 - \frac{C_{R_{(k)}}}{C_p} \tag{2.40}$$

where:

C_p: the Peukert capacity, considered constant (Ah).

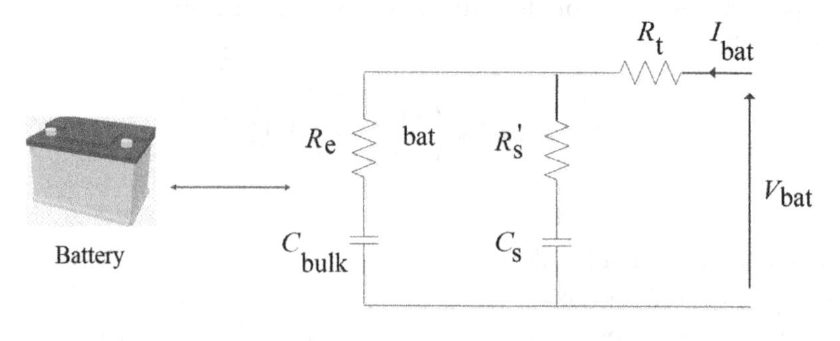

Figure 2.7 Equivalent circuit of a lead–acid battery.

2.3.1.3 Chopper

The connection of a PV generator to a load requires an adaptation system, to ensure the operation at the MPP. This consists in varying the duty cycle of the DC–DC converter (chopper), which is interposed between the PVP and the load. Fig. 2.8 shows the block diagram of the DC–DC adaptation of the PV generator to a load (Ben Ammar, 2011; Yahyaoui et al., 2016).

The variation of the duty cycle used to control the chopper is performed by the MPPT algorithm (Yahyaoui, Chaabene, & Tadeo, 2016). Hence, in this book, three type of DC-DC converters are cited: the buck for applications that need to decrease the PV voltage, the boost for applications that require increasing the PV voltage, and the buck–boost if the applications require operating in the two modes, buck and boost. In the application proposed in this book, since the MPP voltage of the PVPs is higher than the battery bank voltage, it is obvious then to use the buck chopper (Fig. 2.9). This DC–DC converter, characterized by low electric energy consumption and a high efficiency, comprises inductors, capacitors, and electronic switches. For the buck choppers, the electronic switch more used is the MOSFET fast switch (Ben Ammar, 2011; Salimi, Soltani, Markadeh, & Abjadi, 2013).

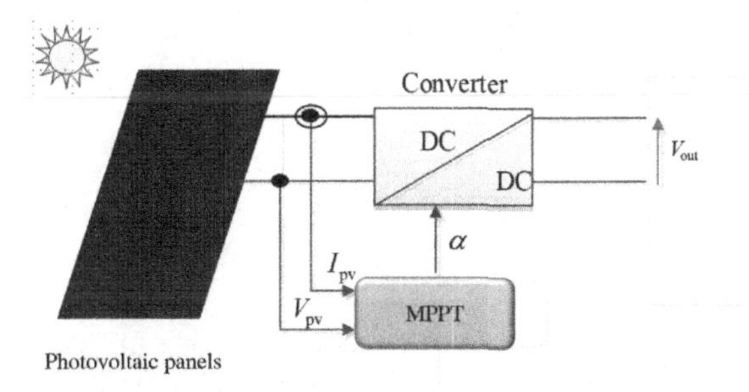

Figure 2.8 DC–DC adaptation of the PV generator to a load.

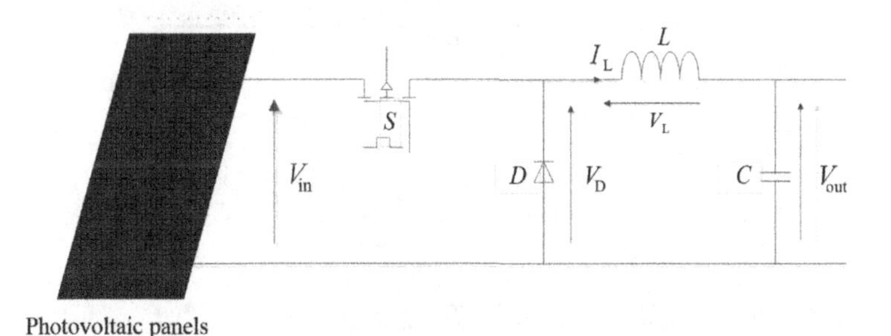

Figure 2.9 Principle circuit of the buck chopper.

Function principle

The adaptation of the load with the solar panel is based in fixing the average voltage V_{out} at the output of the converter that meets the following criteria (Ben Ammar, 2011; Yahyaoui et al., 2016):

- $V_{out} \prec V_{in}$.
- V_{out} is adjustable in the desired range.

The operation of the chopper is deduced from the switch S behavior analysis. In fact, two operating phases can be distinguished (Ben Ammar, 2011):

- When the switch S is *on* during $0 \prec t \prec \alpha T$, the diode is reverse biased ($V_D = -V_{in}$) and $V_L = V_{in} - V_{out}$.
- When the switch S is *off* during $\alpha T \prec t \prec T$, then the diode is directly biased ($V_D = 0$) and $V_L = -V_{out}$.

Two operating modes can be distinguished, following the current $I_L(t)$ in the inductance L:

Continuous operating mode During the continuous mode, the current $I_L(t)$ never reaches zero. The operating diagram of the buck chopper in the continuous mode is described by Fig. 2.10. The variation of $I_L(t)$ is given by (Ben Ammar, 2011; Reis, Miranda, Lemes, Viajante, & Chaves, 2015; Salimi et al., 2013):

$$V_L(t) = L\frac{dI_L(t)}{dt} \tag{2.41}$$

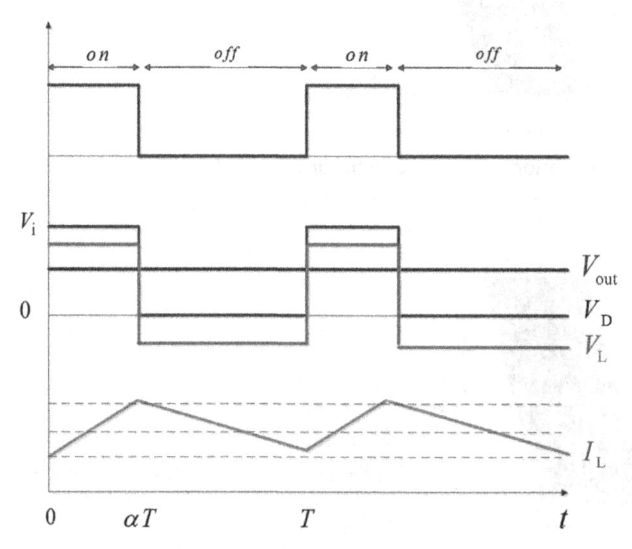

Figure 2.10 Ideal wave forms for the buck converter operating in the continuous mode.

- When $0 \prec t \prec \alpha T$, the MOSFET S is saturated and the current $I_L(t)$ increases:

$$\Delta I_{L\ \mathrm{on}}(t) = \int_0^{\alpha T} \frac{V_{pv} - V_{out}}{L} dt = \frac{V_{pv} - V_{out}}{L} \alpha T \tag{2.42}$$

- When $\alpha T \prec t \prec T$, the MOSFET S is blocked and $I_L(t)$ decreases:

$$\Delta I_{L\ \mathrm{off}}(t) = \int_{\alpha T}^{T} \frac{-V_{out}}{L} dt = \frac{-V_{out}}{L} (T - \alpha T) \tag{2.43}$$

The current in the inductance when S is *on* or *off* is the same. Hence:

$$V_{out} = \alpha\ V_{pv} \tag{2.44}$$

Discontinuous operating mode During this mode, the current $I_L(t)$ reaches zero when the MOSFET is blocked (Fig. 2.11). The energy stocked in the inductance is zero. Using the principle given by Eq. (2.44), the following relation is obtained (Ben Ammar, 2011; Riggio & Houghton, 2014):

$$(V_{pv} - V_{out})\alpha T - V_{out}\delta T = 0 \tag{2.45}$$

Thus:

$$\delta = \frac{V_{pv} - V_{out}}{V_{out}} \alpha \tag{2.46}$$

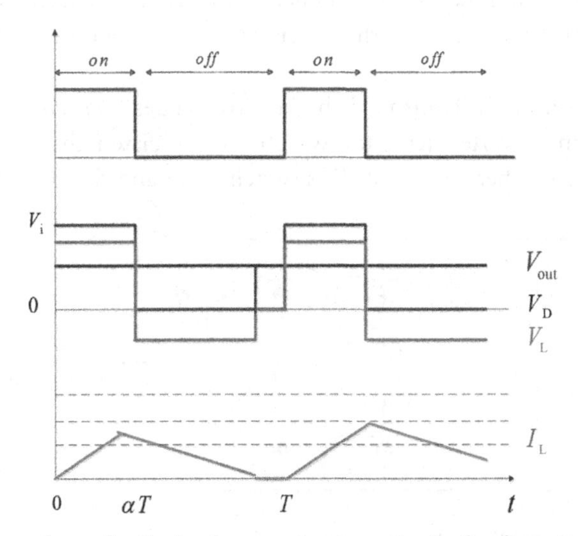

Figure 2.11 Ideal wave forms for the buck converter operating in the discontinuous mode.

The value of the capacity C is important that the mean value of the current in the capacity is zero. Hence (Ben Ammar, 2011; Riggio & Houghton, 2014):

$$\bar{I}_L = I_{\text{out}} \tag{2.47}$$

Using Fig. 2.11, the mean value of the inductance current $\bar{I}_L$ is given by (Ben Ammar, 2011; Riggio & Houghton, 2014):

$$\bar{I}_L = \frac{I_{L\ \max}(\alpha + \delta)}{2} \tag{2.48}$$

Knowing that:

$$I_{L\ max} = \frac{V_{\text{pv}} - V_{\text{out}}}{L} \alpha T \tag{2.49}$$

The output voltage is expressed by (Ben Ammar, 2011; Riggio & Houghton, 2014):

$$V_{out} = V_{\text{pv}} \frac{1}{\frac{2LI_{\text{out}}}{\alpha^2 V_{\text{pv}} T} + 1} \tag{2.50}$$

2.3.1.4 Inverter

As in most research related to water pumping, the motor pump adopted is an IM, since it is simple in control and its price is encouraging (Sallem et al., 2009a, 2009b). Hence, since the IM needs to be supplied by AC signals, a three-phase inverter is used to convert the signals from DC to AC. Indeed, the inverter is composed of six IGBT switches each shunted in antiparallel by a fast freewheeling diode, in order to return the negative current to the filter capacitor provided at the input of the converter (Fig. 2.12).

Hence, the inverter is controlled by analog values. T_i and T_i' are the ideal switches of the same inverter arm, for which are associated the logic control signals S_i and $\overline{S}_i$, respectively, where $S_i = 1$ if T_i is switched on and $S_i = 0$ if T_i is switched off.

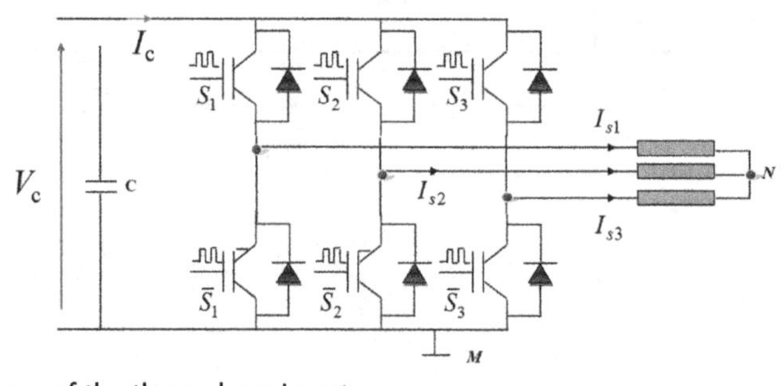

Figure 2.12 Schema of the three-phase inverter.

Thus, the composed and simple voltages vectors, and the currents vector, which depends on the control signals and the input voltage V_c are used to define the composed voltage vectors U_{s1s2}, U_{s2s3}, and U_{s3s1} relative to the common point N of the load or the ground M as follows (Eq. 2.51) (Ben Ammar, 2011; Miveh, Rahmat, Ghadimi, & Mustafa, 2016):

$$\begin{cases} V_{s1} - V_{s2} = V_{s1M} - V_{s2M} = V_{s1N} - V_{s2N} \\ V_{s2} - V_{s3} = V_{s2M} - V_{s3M} = V_{s2N} - V_{s3N} \\ V_{s3} - V_{s1} = V_{s3M} - V_{s1M} = V_{s3N} - V_{s1N} \end{cases} \tag{2.51}$$

Since $V_{iM} = V_c S_{i/i=s1,s2,s3}$ and the load is balanced, the stator voltage ensures that $V_{s1} + V_{s2} + V_{s3} = 0$. Hence (Miveh et al., 2016):

$$\begin{bmatrix} V_{s1} \\ V_{s2} \\ V_{s3} \end{bmatrix} = \frac{V_c}{3} \begin{bmatrix} 2 & -1 & -1 \\ -1 & 2 & -1 \\ -1 & -1 & 2 \end{bmatrix} \begin{bmatrix} S_{s1} \\ S_{s2} \\ S_{s2} \end{bmatrix} \tag{2.52}$$

The input current I_c can be expressed by Eq. (2.53) (Miveh et al., 2016):

$$I_c = S_{s1} I_{s1} + S_{s2} I_{s2} + S_{s3} I_{s3} \tag{2.53}$$

The sinusoidal pulse-width modulation (PWM) allows the control signals S_{s1}, S_{s2}, and S_{s3} to be deduced. In fact, this modulation obtained (Altin & Ozdemir, 2013; Ben Ammar, 2011) by comparing a referential sinusoidal signal called modulating wave (characterized by the frequency f_r and the amplitude V_r) with a triangular signal called carrier wave, characterized by a frequency $f_p \succ \succ f_r$, and the amplitude V_p (Altin & Ozdemir, 2013; Miveh et al., 2016). When the two signals take the same value, the IGBTs change the state, so PWM signals are generated, with the frequency f_p.

2.3.1.5 Pump

A pump uses the electric power (in the present studied case, it is given by the panel and/or the battery) to provide mechanical energy to the water (Fig. 2.1) (Appendix B). In the literature, these pumps are either positive displacement or dynamic pumps (Casoli & Anthony, 2013; Yin, Lin, Li, Liu, & Gu, 2015). Positive displacement pumps are used in applications characterized by a constant discharge speed, or at high heads and low flow rates, since this type of pump delivers periodic flows. In the application presented in this book, dynamic pumps are used, as they are adequate when the application needs a variable discharge speed, or at high flow rate and low or moderate heads (Casoli & Anthony, 2013; Roger, Perez, Campana, Castiel, & Dupuy, 1978).

In this context, centrifugal pumps are commonly used, since they require less torque to start, and produce more head than other dynamic pumps at a variable speed (Roger et al., 1978). Moreover, in addition to their simplicity and low cost, they are characterized by their low maintenance; moreover centrifugal pumps are available for different flow rates and heads (Roger et al., 1978). Hence, for this application, a centrifugal submerged pump is selected.

As it has been previously mentioned, the centrifuge pump is supplied by an IM (Appendix C). To model it, the vector transformation has been used, and gives the following dynamic model of the IM in a (d, q) frame (Eq. 2.54) (Sallem et al., 2009a, 2009b):

$$\begin{cases} v_{sd} = R_{ss}I_{sd} + L_s\frac{d}{dt}I_{sd} + m\frac{d}{dt}I_{rd} - w_s(L_sI_{sq} + mI_{rq}) \\ v_{sq} = R_{ss}I_{sq} + L_s\frac{d}{dt}I_{sq} + m\frac{d}{dt}I_{rq} + w_s(L_sI_{sd} + mI_{rd}) \\ 0 = R_{rr}I_{rd} + L_r\frac{d}{dt}I_{rd} + m\frac{d}{dt}I_{sd} - w_g(L_rI_{rq} + mI_{sq}) \\ 0 = R_{rr}I_{rq} + L_r\frac{d}{dt}I_{rq} + m\frac{d}{dt}I_{sq} + w_g(L_rI_{rd} + mI_{sd}) \end{cases} \tag{2.54}$$

with:

v_{sd}: the stator voltage in the direct axe (V),
v_{sq}: the stator voltage in the quadrature axe (V),
I_{sd}: the stator current in the direct axe (A),
I_{sq}: the stator current in the quadrature axe (A),
I_{rd}: the rotor current in the direct axe (A),
I_{rq}: the rotor current in the quadrature axe (A),
R_{ss}: the stator resistance per phase (Ω),
R_{rr}: the rotor resistance per phase (Ω),
L_s: the cyclic stator inductance per phase (H),
L_r: the cyclic rotor inductance per phase (H),
m: the mutual inductance stator–rotor (H),
w_g: the rotor pulsations (rad/s).

The electromagnetic torque C_{em} is given by Eq. (2.55) (Sallem et al., 2009a, 2009b; Yahyaoui, Sallem, Chaabene, & Tadeo, 2012):

$$C_{em} = p\frac{m}{L_r}(\varphi_{rd}I_{sq} - \varphi_{rq}I_{sd}) \tag{2.55}$$

where:

p: the number of poles pairs,
φ_{rd}: the rotor flux in the direct axe,
φ_{rq}: the rotor flux in the quadrature axe.

The mechanical equation is as follows (Eq. (2.56)) (Sallem et al., 2009a, 2009b; Yahyaoui et al., 2012):

$$\frac{\mathrm{d}}{\mathrm{d}t} w_{\mathrm{m}} = \frac{1}{J} p(C_{\mathrm{em}} - C_{\mathrm{r}}) \tag{2.56}$$

Here, the IM is coupled to a centrifuge pump whose torque is given by Eq. (2.57) (Sallem et al., 2009a, 2009b; Yahyaoui et al., 2012):

$$C_{\mathrm{r}} = k w_{\mathrm{m}} \tag{2.57}$$

where:

$$k = \frac{C_{\mathrm{em\ max}}}{w_{\mathrm{m\ max}}} \tag{2.58}$$

where:

$C_{\mathrm{em\ max}}$ is the maximum torque and $w_{m\ max}$ is the maximum speed.

During the pump operating, the rotor flux is positioned in a privileged position ($\varphi_{\mathrm{rd}} = \varphi_{\mathrm{r}}$ and $\varphi_{\mathrm{rq}} = 0$), thanks to the rotor field oriented control (RFOC) (Fig. 2.13) (Yahyaoui et al., 2012). This method consists in controlling, independently, the flux and the current at a constant speed, by acting on the rotational speed w_{m} and the rotor flux φ_{r}, using the direct and the quadrature components of the stator current I_{sd} and I_{sq}, respectively. In this sense, the flux and the current are controlled independently to impose the electromagnetic torque C_{em}. Hence, Eq. (2.55) becomes (Sallem et al., 2009b; Yahyaoui et al., 2012):

$$C_{\mathrm{em}} = p\frac{M}{L_{\mathrm{r}}} \varphi_{\mathrm{r}} I_{\mathrm{sq}} \tag{2.59}$$

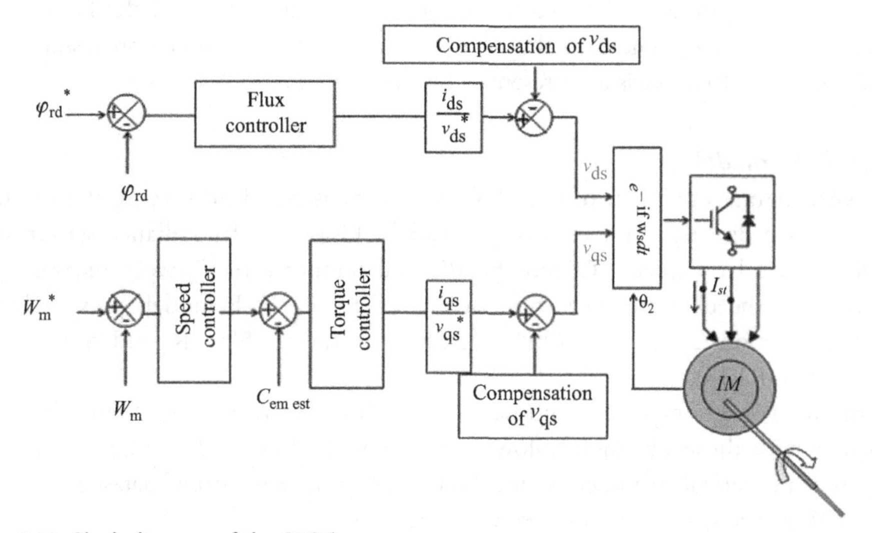

Figure 2.13 Block diagram of the RFOC.

Since the rotor flux is not accessible, it is estimated using the DC component I_{sd} as follows (Sallem et al., 2009b; Yahyaoui et al., 2012):

$$\varphi_{rd} = \frac{M}{1 + \tau_r s} I_{sd} \tag{2.60}$$

In the RFOC, the regulation of the current, flux, and the speed is done via PI regulators. The mechanical power of the IM coupled to the pump is (Sallem et al., 2009b; Yahyaoui et al., 2012):

$$P_L = C_r w_m \tag{2.61}$$

Coupled to the IM, the total mechanical power is expressed by (Sallem et al., 2009b; Yahyaoui et al., 2012):

$$P_L = \frac{V\, g\, \rho\, H_h}{\eta_p\, \Delta t} \tag{2.62}$$

where:

V: the pumped water volume (m^3),
g: the gravity acceleration (m/s^2),
ρ: the water density (kg/m^3),
H_h: the head height (m),
η_p: the pump efficiency,
Δt: the pumping duration (h).

2.3.2 Experimental validation and modeling results

The models validations of the two main components: the PVP and the battery bank are performed and presented in this section. Moreover, the simulation results of the inverter and the IM models are presented below.

2.3.2.1 PVPs models

The experimental validation of two different panels is based on varying the resistance used as a load directly connected to the panels. Using a solar radiance sensor and a PT1000, the solar radiance G and the PV cell temperature T_c were measured and used to draw the characteristics that correspond to the PVP models previously presented in Section 2.3.1, Eqs. (2.20)–(2.28), for the TE500CR and Sunel panels (Figs. 2.14 and 2.15).

Current, voltage, radiation, and panel temperature were measured, and the results compared with those obtained following the models (Eqs. (2.20)–(2.28)) (Fig. 2.3), using the numerical parameters in Table 2.2, obtained from datasheets of the TE500CR and Sunel panels, and from the literature.

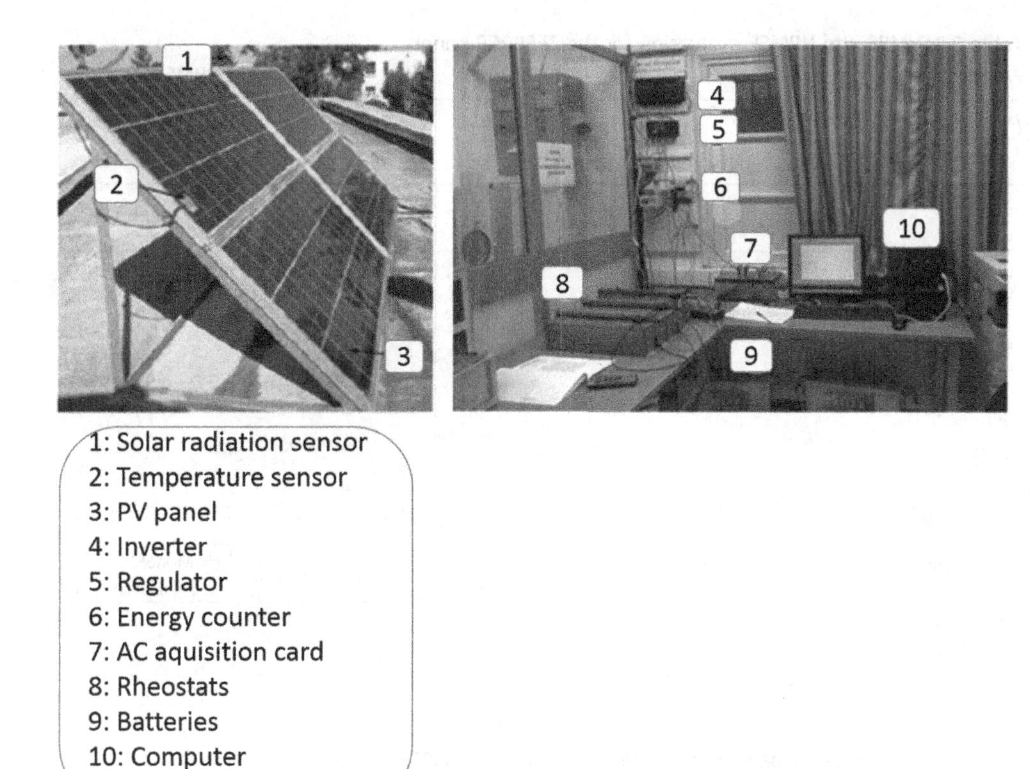

Figure 2.14 Laboratory system used for validation of the TE500CR model.

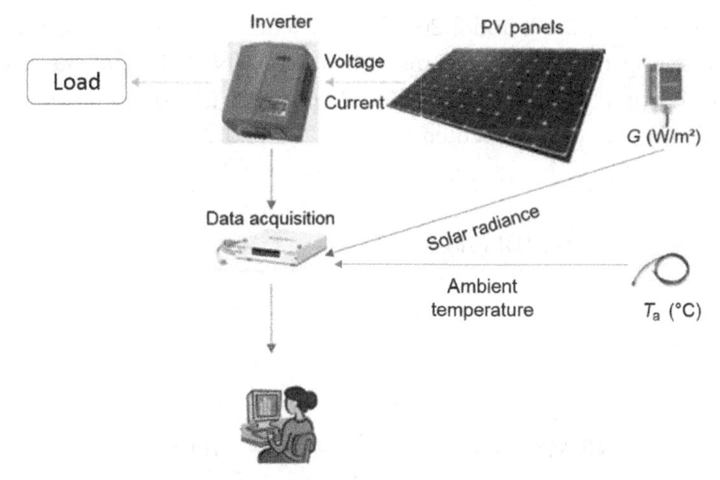

Figure 2.15 Laboratory system used for validation of the Sunel model.

Table 2.2 NMBE and NRMSE evaluation for the TE500CR panel

Parameters	NMBE	NRMSE
P_{pv} (yield model)	3.77%	4.59%
I_{pv} (nonlinear model)	4.55%	6.89%

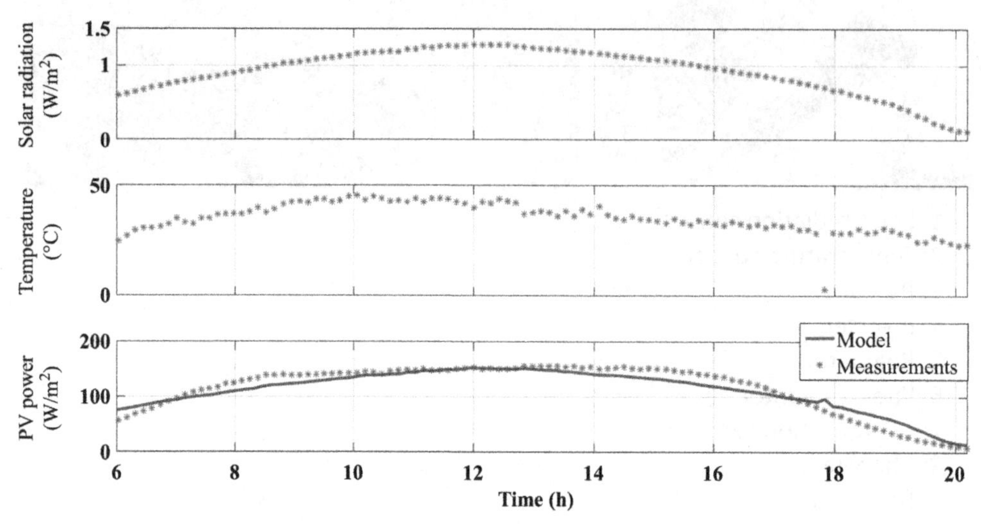

Figure 2.16 Yield-based panel model validation for the TE500CR panel.

The Sunel module has been divided in three substrings connected in serial. Hence, the models validation for one Sunel substring is considered. The experimental characteristics of the yield-based panel model (Eqs. (2.20)–(2.22)) are presented in Figs. 2.16 and 2.18. The characteristics of the nonlinear model (Eqs. (2.23)–(2.28)) are presented in Figs. 2.17, 2.19, and 2.20. The efficiencies of the proposed models are evaluated by calculating the normalized mean bias error (NMBE) and the normalized root-mean-square error (NRMSE), using Eqs. (2.63) and (2.64) (Ben Ammar, 2011; Yahyaoui et al., 2016): They are presented in Tables 2.3 and 2.4.

$$\text{NMBE}(\%) = \frac{\sum_{i=1}^{N} \tilde{X}_i - X_i}{\sum_{i=1}^{N} X_i} * 100 \tag{2.63}$$

$$\text{NRMSE}(\%) = \frac{\sqrt{\frac{1}{N}\sum_{i=1}^{N} (\tilde{X}_i - X_i)^2}}{\frac{1}{N}\sum_{i=1}^{N} X_i} * 100 \tag{2.64}$$

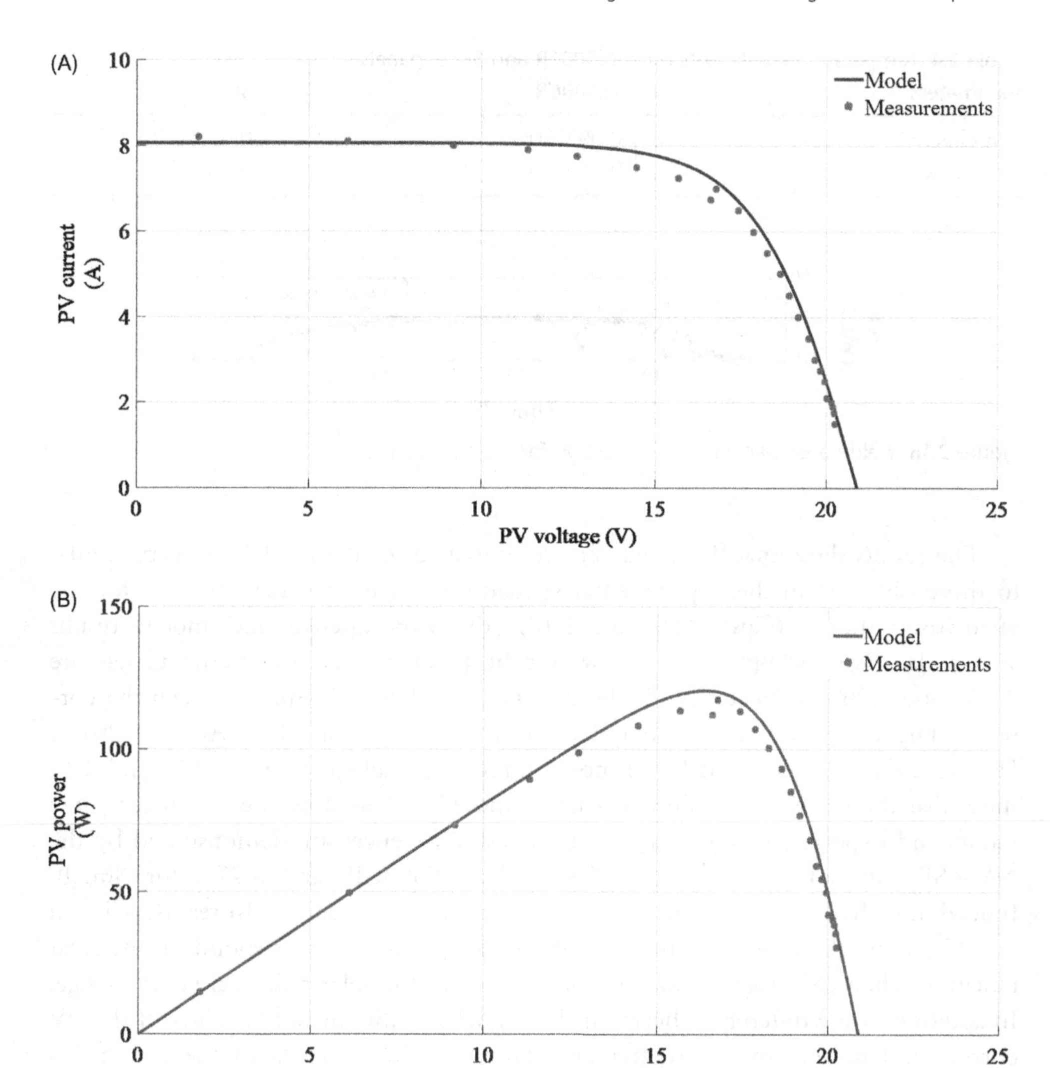

Figure 2.17 TE500CR PVP I-V (A) and P–V (B) curves at $G \approx 480$ W/m^2 and $T_c \approx$ 25°C.

Table 2.3 NMBE and NRMSE evaluation for the Sunel panel

Parameters	NMBE	NRMSE
P_{pv} (yield model)	2.26%	3.45%
I_{pv} (nonlinear model)	−2.19%	5.77%

Table 2.4 Temperature coefficients *a* for TE500CR and Sunel panels

Parameters	TE500CR	Sunel
Datasheet	0.095%/K	0.039%/K
Experimental	0.083%/K	0.06%/K

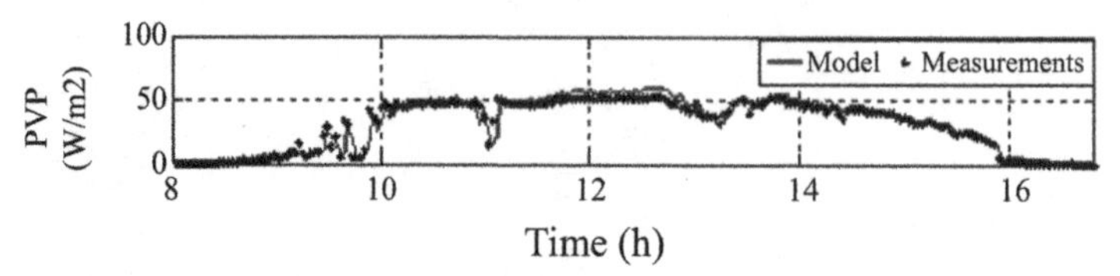

Figure 2.18 Yield-based panel model validation for the Sunel panel.

The results show that the characteristics obtained for the models are very similar to those obtained in the experimental validations, for both panels. In fact, for the yield-based model (Figs. 2.16 and 2.18), the experimental and model results are similar. For example the NMBE for both TE500CR and Sunel panels are 3.77% and 2.26%, respectively (Tables 2.3 and 2.4). The difference between the corresponding curves can be attributed to measurement errors (Ben Ammar, 2011). Hence, the use of this yield-based model for sizing is adequate. Figs. 2.17 and 2.19 show that there are small differences in I_{sc} and V_c obtained by the nonlinear panel model and experiments for both panels. These differences are demonstrated by the NMRSE values for both panels (6.89% for TE500CR and 5.77% for Sunel). Indeed, the change in the solar radiation affects the measurement: In reality, it is not possible to get the exact PV current and voltage values that correspond to one solar radiation when changing the load resistance, due to the solar radiation rapid change. In addition, some differences between the model and the measured values of the PV current and power are due to the uncertainty in the selection of the parameters values and to simplifications adopted when modeling (De Blas et al., 2002). For example, there is a small difference between the values presented in Table 2.5 of the temperature coefficient a of the short circuit current I_{sc} given by the datasheet, and the value experimentally calculated using its definition given by Eq. (2.65) (Ben Ammar, 2011):

$$a = \frac{I_{sc\ T_2} - I_{sc\ T_1}}{I_{sc\ T_1}} * \frac{1}{T_2 - T_1} \tag{2.65}$$

where:

$I_{sc\ T_2}$: the short circuit current at the temperature T_2 (A),
$I_{sc\ T_1}$: the short circuit current at the temperature T_1 (A).

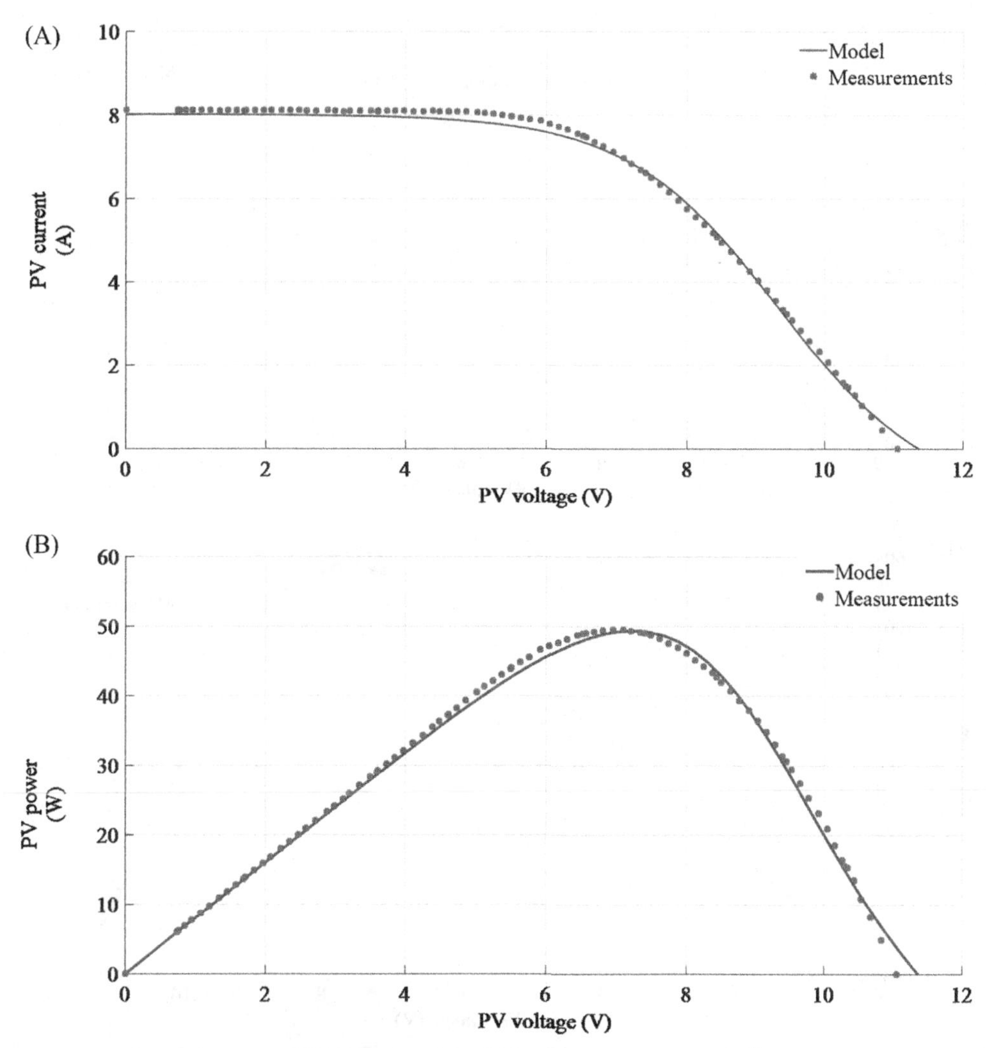

Figure 2.19 Sunel PV substring I–V (A) and P–V (B) curves at $G \approx 864$ W/m^2 and $T_c \approx 45$°C.

Table 2.5 Temperature coefficients *a* for TE500CR and Sunel PV panels

Parameters	TE500CR (%/K)	Sunel (%/K)
Datasheet	0.095	0.039
Experimental	0.083	0.06

Both the model and experimental characteristics of the PV currents show that the increase in the solar radiation G implies an increase in the generated current. Instead, the temperature increase at the module surface T_c decreases the open-circuit voltage V_c (Figs. 2.19 and 2.20).

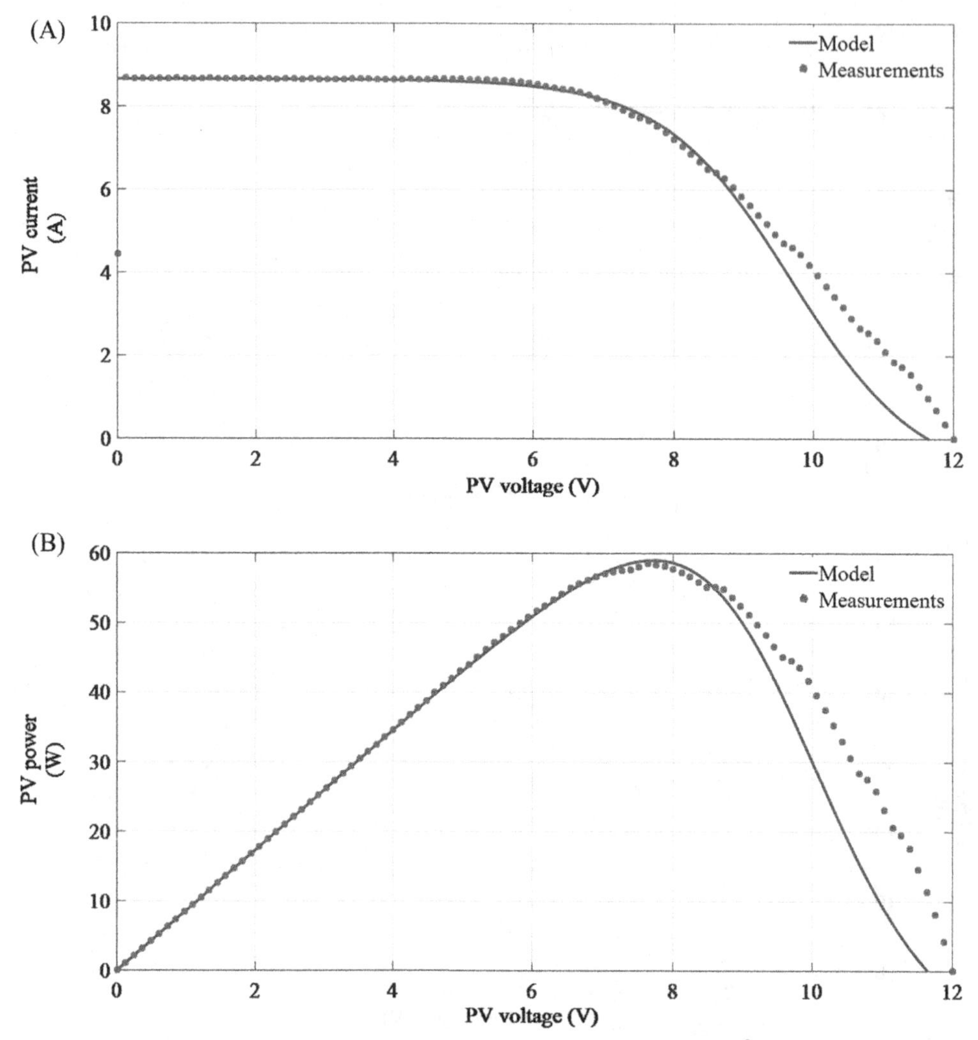

Figure 2.20 Sunel PV substring I—V (A) and P—V (B) curves at $G \approx 950\ W/m^2$ and $T_c \approx 40°C$.

2.3.2.2 MPPT results

In literature, several algorithms for MPPT have been developed and validated (Hohm & Ropp, 2000a, 2000b), for example, the lookup table MPPT (Charfi & Chaabene, 2014), the neuro-fuzzy (Veerachary & Yadaiah, 2000), the incremental conductance (Veerachary & Yadaiah, 2000), and the P&O (Femia et al., 2005) methods. The results of the MPPT using these methods are presented and explained here. In fact, using measured climatic data (G, T_a) of Medjez El Beb (Northern of Tunisia, latitude: 36.39°; longitude: 9.6°) during a typical day in July, the MPPT algorithms have been compared in terms of the PV power P_{mpp}, current I_{mpp}, voltage V_{mpp}, and the duty cycle α deviations. The performance indexes are expressed by the NMBE and the NRMSE, given by Eqs. (2.63) and (2.64) (Charfi & Chaabene, 2014). The results comparison of the studied MPPT methods is presented in Figs. 2.21–2.24 and summarized in Tables 2.6–2.8.

The obtained results show that the neuro-fuzzy MPPT method presents the highest NRMSE error for the power, current, voltage, and the duty cycle. This is due to the need of a continuous update for the database used for the data training. Although updated data are used for the Smart MPPT, its NMBE and NRMSE errors are more important than those of the incremental conductance or the P&O methods. This is because Smart MPPT uses the minimum values of G and T_a intervals, which makes the working point different from the real MPP.

The incremental conductance and the P&O methods present similar results. For instance, the NRMSE for the incremental conductance and the P&O are,

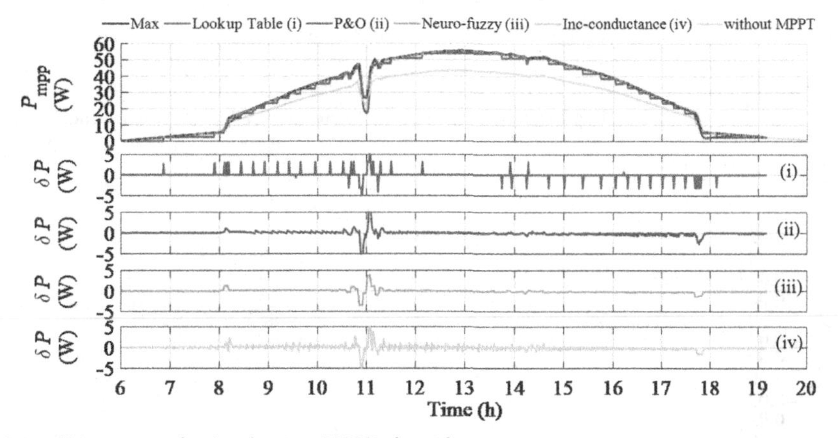

Figure 2.21 PV powers obtained using MPPT algorithms.

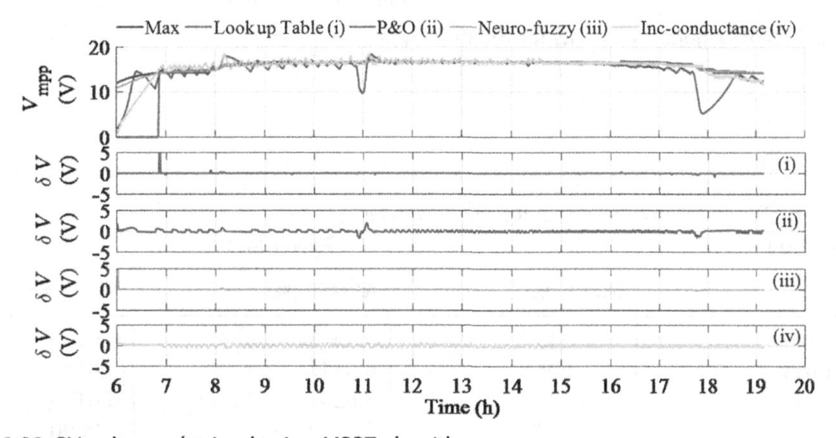

Figure 2.22 PV voltage obtained using MPPT algorithms.

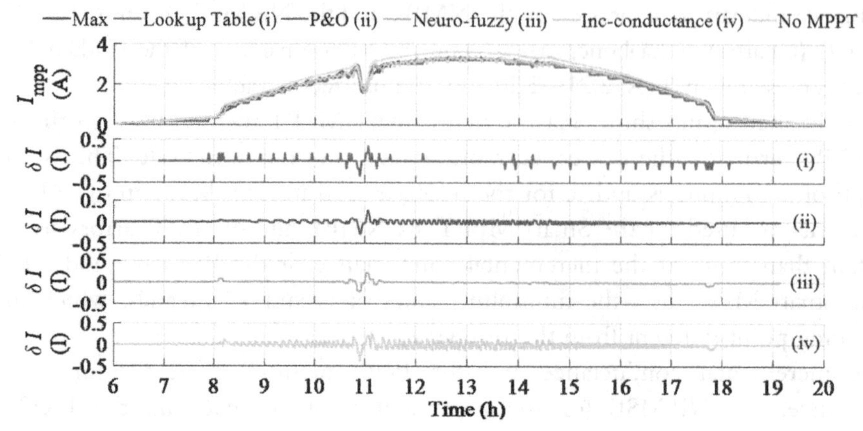

Figure 2.23 PV current obtained using MPPT algorithms.

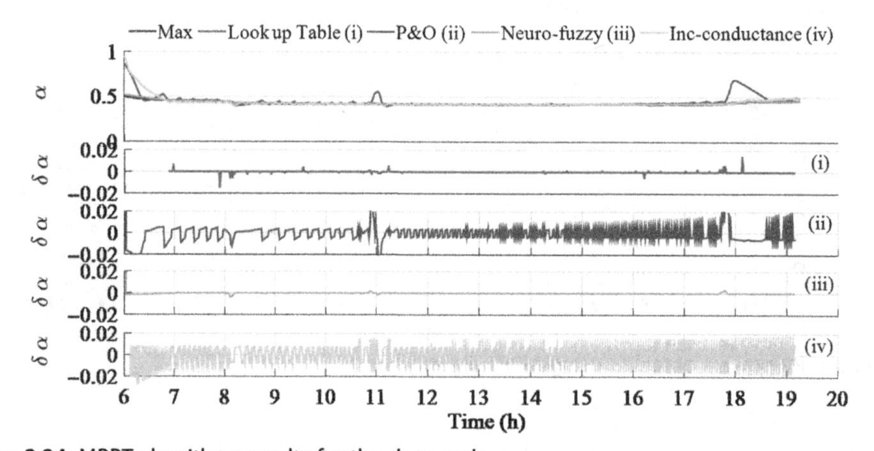

Figure 2.24 MPPT algorithms results for the duty cycle α.

Table 2.6 MPPT algorithms evaluation for the PV power

MPPT method	Power loss		
	NMBE (%)	NRMSE (%)	P_{mpp} variation (%)
Lookup Table	− 3.8437	4.9311	0.0109
Neuro-fuzzy	− 1.0984	6.5541	0.0101
Inc-cond	− 0.4235	1.6241	0.0106
P&O	− 1.0725	3.6982	0.0096

Table 2.7 MPPT algorithms evaluation for PV voltage

MPPT method	**Voltage loss**		
	NMBE	**NRMSE**	**V_{mpp} variation**
Lookup table	0.2592%	1.2872%	0.1174%
Neuro-fuzzy	3.1273%	39.7398%	0.1002%
Incremental conductance	−2.6232%	11.2374%	0.1292%
P&O	−4.8122%	13.4396%	0.0917%

Table 2.8 MPPT algorithms evaluation for PV current

MPPT method	**Current loss**		
	NMBE	**NRMSE**	**I_{mpp} variation**
Lookup table	−4.6251%	5.5891%	0.0130%
Neuro-fuzzy	−2.9987%	33.2998%	0.0136%
Incremental conductance	−0.0532%	3.3358%	0.0139%
P&O	0.7172%	3.2213%	0.0117%

respectively, 1.6241% and 3.6982% for the power; 3.3358% and 3.2213% for the current. Hence, the errors values are close. P&O MPPT method is easy in implementation, and characterized by a low cost in installation, compared of the incremental conductance (Oi, 2005b; Taufik EE410 Power Electronics I, 2004). Since these errors will not cause a great difference for the power and current at the MPP, the P&O is chose in the present application to track the MPP.

2.3.2.3 Battery bank model

The validation of the battery model is carried out by keeping the battery voltage constant (using the voltage regulator) and varying the load resistance. The results, presented in Fig. 2.25, are compared to those obtained by the model detailed in Section 2.3.1, using data given by the datasheet and the literature, shown in Table 2.9.

The *dod* values obtained by the experimental validation (directly obtained from the battery regulator) and the battery bank model are similar. This proves that the adopted model is efficient. Moreover, the relation between the *dod* and the battery current is clear: when the battery is in charge, I_{bat} is positive and the *dod* decreases. For example, using the measured results, starting from 50% of the battery capacity, the battery is charged for 10 min with a constant current equal to 4.5 A, and the *dod* decreases to 16%. When the battery is discharging with a constant current equal to 9.7 A for 10 min, the *dod* increases from 16% to 68%.

2.3.2.4 Chopper model

Here, a buck converter is designed using the parameters provided by the datasheet (Table 2.10).

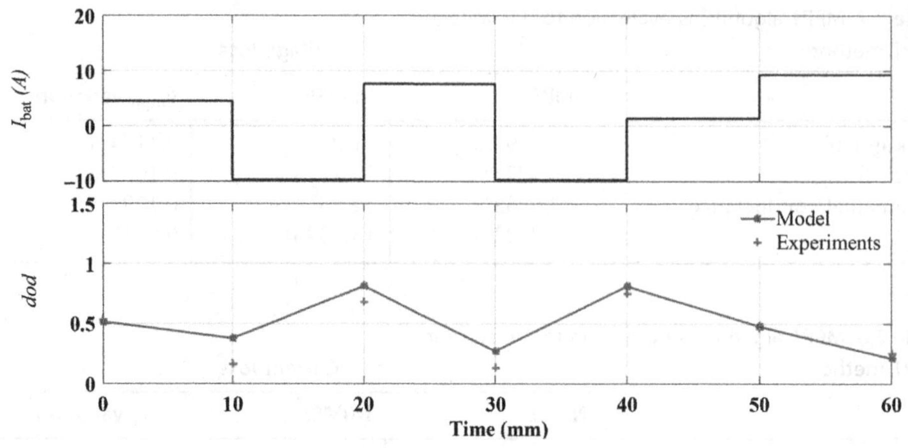

Figure 2.25 Experimental validation results for the battery model.

Table 2.9 Numerical parameters for the lead–acid battery

Parameters	Values
C_p	210 Ah
V_{bat}	12 V
R_t	0.0126 Ω (Ben Salah & Ouali, 2012)
R_e	0.0168 Ω (Ben Salah & Ouali, 2012)
R_s'	0.0168 Ω (Ben Salah & Ouali, 2012)
k_p	1.12 (Chaabene, 2009)

Table 2.10 Design specification for the buck converter

Specifications	Values
Input voltage	0–20 (V)
Input current	0–4.5 (A)
Output voltage	12 V
Output current	5 A
Maximum output power	60 W
Switching frequency	50 KHz
Duty cycle	$0.1 < \alpha < 0.5$

Inductor selection

The selection of the inductor size depends on the rate of change in the inductor current. In fact, less than 5% in the current ripple is permitted. This current variation is expressed as follows (Oi, 2005b):

$$\Delta i_L = \frac{V_{pv}\alpha}{Lf} \tag{2.66}$$

where:

V_{pv}: the PV voltage (V),

α: the duty cycle,

f: the switching frequency (Hz).

Hence, the inductance value can be deduced by:

$$L = \frac{V_{pv}\alpha}{\Delta i_L f} \tag{2.67}$$

Capacitor selection

The design criterion for the capacitor is that the ripple voltage across it should be less than 5%. The average voltage across the capacitor C is given by Eq. (2.68) (Oi, 2005b):

$$\Delta V_C = 0.05(V_{pv} + V_{out}) \tag{2.68}$$

The value of the capacity C is calculated with the following equation (Oi, 2005b):

$$C = \frac{V_{out}\alpha}{Rf\Delta V_C} \tag{2.69}$$

where R is the equivalent load resistance which is given by (Oi, 2005b):

$$R = \frac{V_{out}^2}{P_{out}} \tag{2.70}$$

Diode selection

Schottky diode is selected since it has a low forward voltage and a good reverse recovery time (typically 5–10 ns) (Taufik EE410 Power Electronics I, 2004). The peak reverse voltage V_{rrm} of the diode is the same as the voltage of the capacity C. Generally a 30% of safety factor is used. The average diode forward current I_f is the same as the output current. Hence, adding 30% of safety factor gives the suitable I_f.

Switch selection

Power-MOSFETs are used in low or medium power applications. The peak voltage of the switch is obtained by the Kirchhoff's Voltage Law the Kirchhoff's Voltage Law (KVL) on the circuit of Fig. 2.26.

$$V_{SW} = V_{pv} - \frac{dI_L}{dt} \tag{2.71}$$

The voltage of the switch SW reaches 20 V. Adding 30% of safety factor gives the suitable voltage for the SW. The peak current is the same as for the diode.

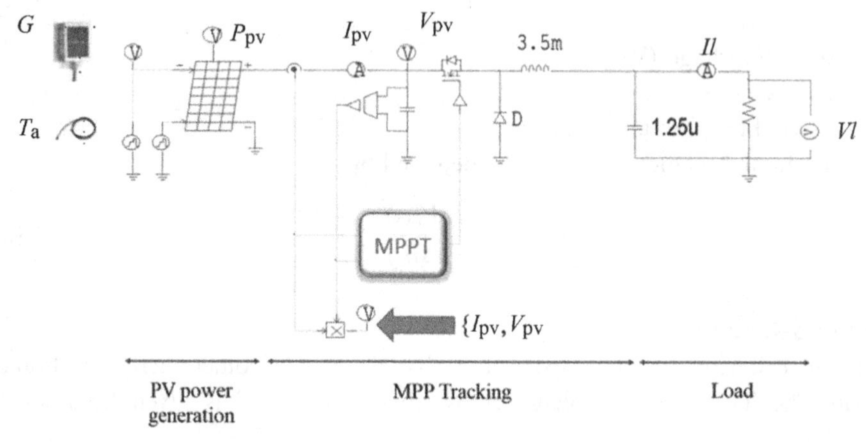

Figure 2.26 Simulation schema of the buck chopper.

Simulation results

The buck chopper is tested using a detailed PowerSim electric simulation using the P&O MPPT. Hence, a buck converter (60 W with a switching frequency of 50 kHz, an inductor value of 3.5 mH and an output filter capacitor of 1.5 μF) is designed (Fig. 2.26).

The PV current and voltage at MPP are tested, using measurements of G and T_a from the target area, to evaluate the performance of P&O. Thus, the duty cycle α, used to generate the PWM signal for the MOSFET switching, is evaluated. The simulation results are given in Fig. 2.27.

Using the buck converter, the obtained results of MPPT show that the MPP is rapidly located using the P&O MPPT method. However, some oscillations in the current, voltage, and power do exist (Fig. 2.27i and ii). This is because the P&O always allows a working point around the real MPP.

2.3.2.5 Inverter model

The PWM signals for the control of the switches T_1, T_2, and T_3 are presented in Fig. 2.28. The results of Figs. 2.29 and 2.30 show that the inverter generates the voltage and current signals with values $V_{eff} = 230\ V$ and $I_{seff} = 14.14\ A$, which corresponds to the nominal values for the IM used.

2.3.2.6 Pump

The simulation results of the IM used for water pumping are presented in Fig. 2.31 using the parameters of Table 2.11: It can be seen that the results reproduce the response expected for the application.

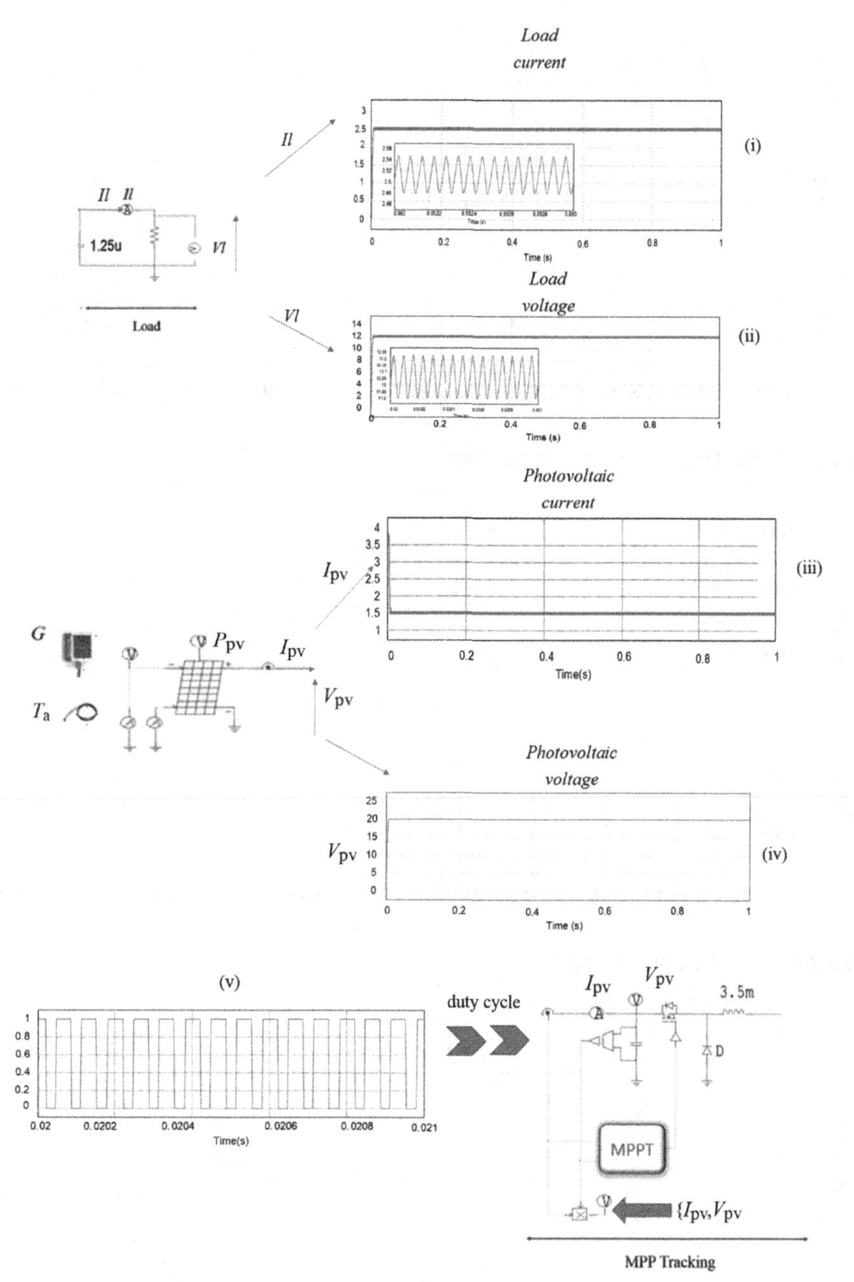

Figure 2.27 Results simulation for $G = 1000\ \text{W/m}^2$, $T_a = 25°\text{C}$: (i) load current, (ii) load voltage, (iii) PV current, (iv) PV voltage, and (v) duty cycle.

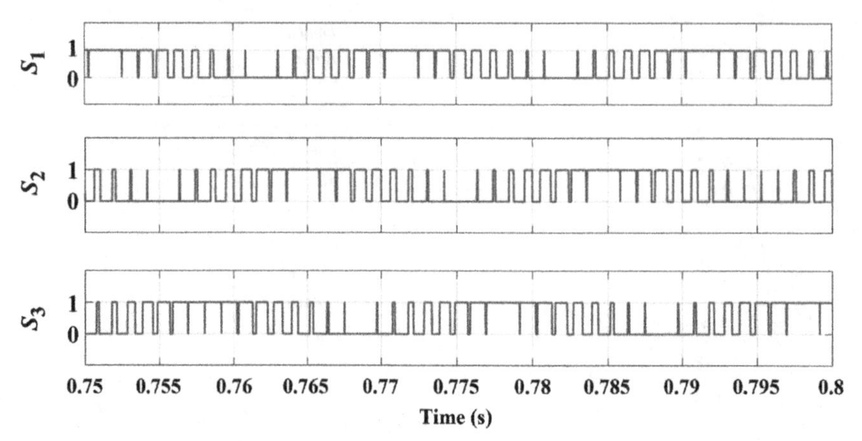

Figure 2.28 Control signals of the inverter switchers.

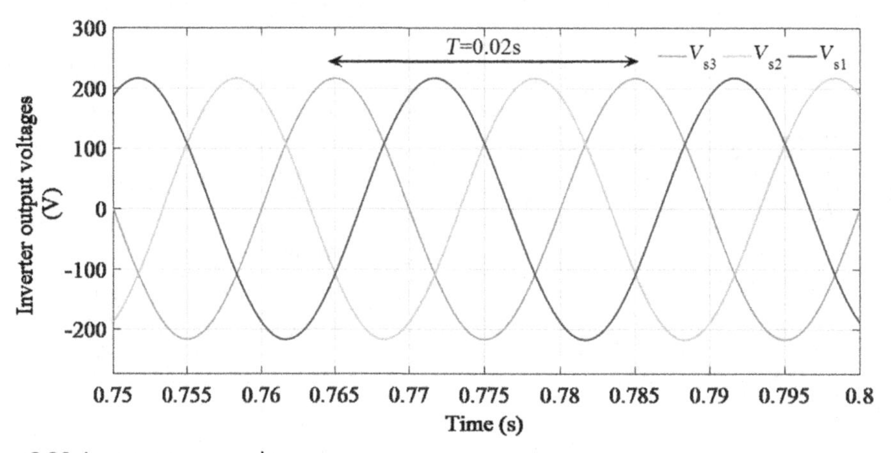

Figure 2.29 Inverter output voltages.

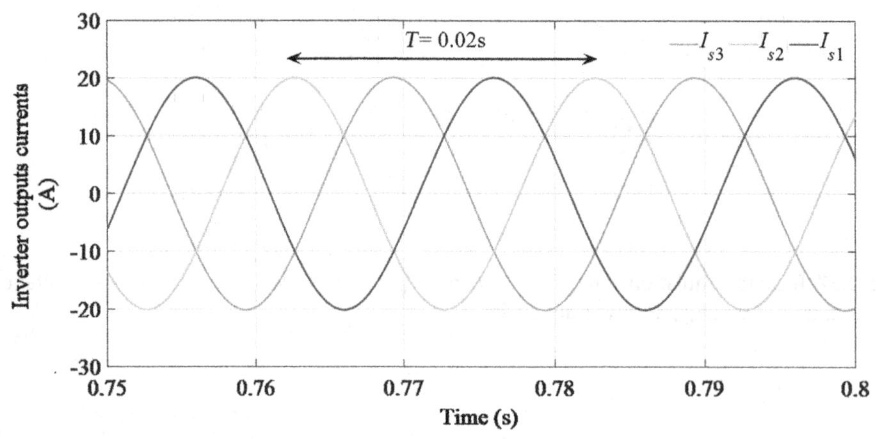

Figure 2.30 Inverter output currents.

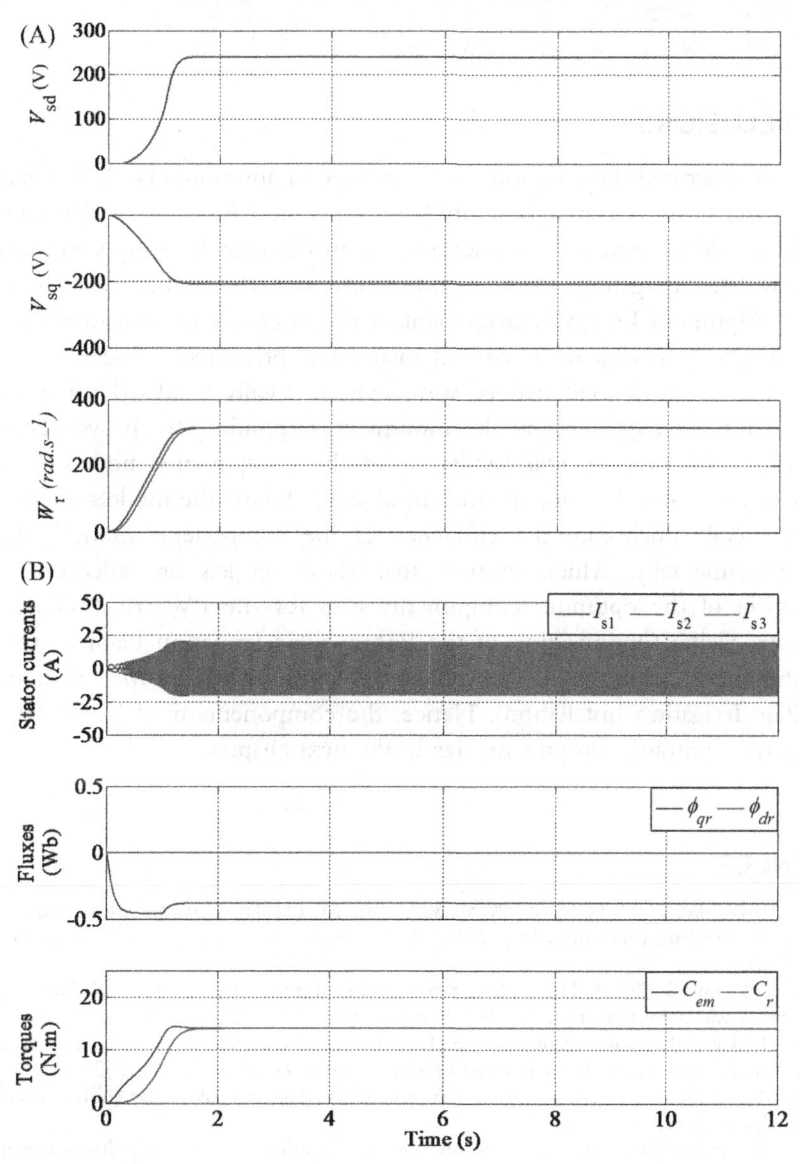

Figure 2.31 IM simulation results using RFOC.

Table 2.11 IM parameters (Sallem et al., 2009a)

Parameters	Values
R_{ss}	5.72 Ω
R_{rr}	4.2 Ω
L_s	0.462 H
L_r	0.462 H
M	0.44 H
P	2
J	0.0049 kg · m^2

2.4 CONCLUSIONS

This chapter described the components models of an autonomous PV installation for pumping water to irrigate an agriculture land. First, models of the installation components, which will be used for the system sizing in Chapter 3, Sizing Optimization of the Photovoltaic Irrigation Plant Components and those that will be used in Chapter 4, Optimum Energy Management of the Photovoltaic Irrigation Installation for designing the management system have been presented. Models for the PV generator and the lead–acid battery were experimentally validated using meteorological data for the target area in the months corresponding to the vegetative cycle of the crop. The experimental validation of the components models proved the efficiency of these models, since the measured values follow the models results.

As a general conclusion, the efficiency of the components models have been proved experimentally, which ensures that these models are adequate for the determination of the optimum components sizes for the PV irrigation installation (see chapter: Sizing Optimization of the Photovoltaic Irrigation Plant Components) and for the energy management (see chapter: Optimum Energy Management of the Photovoltaic Irrigation Installation). Hence, the components models will be used to determine the optimum components size in the next chapter.

REFERENCES

Adamo, F., Attivissimo, F., Di Nisio, A., & Spadavecchia, M. (2011). Characterization and testing of a tool for photovoltaic panel modeling. *IEEE Transactions on Instrumentation and Measurement, 60*(5), 1613–1622.

Altin, N., & Ozdemir, S. (2013). Three-phase three-level grid interactive inverter with fuzzy logic based maximum power point tracking controller. *Energy Conversion and Management, 69*, 17–26.

Arunkumar, T., Jayaprakash, R., Denkenberger, D., Ahsan, A., Okundamiya, M. S., Tanaka, H., et al. (2012). An experimental study on a hemispherical solar still. *Desalination, 286*, 342–348.

Balghouthi, M., Chahbani, M. H., & Guizani, A. (2012). Investigation of a solar cooling installation in Tunisia. *Applied Energy, 98*, 138–148.

Ben Ammar, M. (2011). Contribution à l'optimisation de la gestion des systèmes multi-sources d'énergies renouvelables (Thesis). Tunisia: National Engineering School of Sfax (ENIS).

Ben Salah, C., & Ouali, M. (2012). Energy management of a hybrid photovoltaic system. *International Journal of Energy Research, 36*(1), 130–138.

Bernal-Agustín, J. L., & Dufo-López, R. (2009). Simulation and optimization of stand-alone hybrid renewable energy systems. *Renewable and Sustainable Energy Reviews, 13*(8), 2111–2118.

Bernstein, L., & Francois, L. E. (1973). Comparison of drip, furrow and sprinkler irrigation. *Soil Science, 115*(1), 73–86.

Bouadila, S., Lazaar, M., Skouri, S., Kooli, S., & Farhat, A. (2014). Assessment of the greenhouse climate with a new packed-bed solar air heater at night, in Tunisia. *Renewable and Sustainable Energy Reviews, 35*, 31–41.

Casoli, P., & Anthony, A. (2013). Gray box modeling of an excavator's variable displacement hydraulic pump for fast simulation of excavation cycles. *Control Engineering Practice, 21*(4), 483–494.

Chaabene, M. (2009). "Gestion énergétique des systèmes photovoltaïques" (Master course). Tunisia: National School of Engineering of Sfax.

Charfi, S., &Chaabene, M. (2014). A comparative study of MPPT techniques for PV systems. *IEEE proceedings of the international renewable energy congress (IREC)* (pp. 22–28).

Chenni, R., Makhlouf, M., Kerbache, T., & Bouzid, A. (2007). A detailed modeling method for photovoltaic cells. *Energy, 32*(9), 1724–1730.

Clean Energy Decision Support Centre (2001–2004). *"Photovoltaic project analysis, Chapter", Catalogue no: M39-99/2003E.* RETScreen International, ISBN 0-662-35672-1.

Collares-Pereira, M., & Rabl, A. (1979). The average distribution of solar radiation-correlations between diffuse and hemispherical and between daily and hourly insolation values. *Solar Energy, 22*(2), 155–164.

Collins, J., Kear, G., Li, X., Low, J. C. T., Pletcher, D., Tangirala, R., et al. (2010). A novel flow battery: A lead acid battery based on an electrolyte with soluble lead (II) Part VIII. The cycling of a 10 cm × 10 cm flow cell. *Journal of Power Sources, 195*(6), 1731–1738.

De Blas, M. A., Torres, J. L., Prieto, E., & Garcia, A. M. M. (2002). Selecting a suitable model for characterizing photovoltaic devices. *Renewable Energy, 25*(3), 371–380.

Desai, H.P., & Patel, H.K. (2007). Maximum power point algorithm in PV generation: An overview. *Proceedings of the IEEE conference on power electronics and drive systems (PEDS'07)* (pp. 624–630).

Duffie, J. A., & Beckman, W. A. (2013). *Solar engineering of thermal processes.* John Wiley & Sons, ISBN 978-1-118-43348-5.

El-Sebaii, A. A., Al-Hazmi, F. S., Al-Ghamdi, A. A., & Yaghmour, S. J. (2010). Global, direct and diffuse solar radiation on horizontal and tilted surfaces in Jeddah, Saudi Arabia. *Applied Energy, 87*(2), 568–576.

Erbs, D. G., Klein, S. A., & Duffie, J. A. (1982). Estimation of the diffuse radiation fraction for hourly, daily and monthly-average global radiation. *Solar Energy, 28*(4), 293–302.

Esram, T., & Chapman, P. L. (2007). Comparison of photovoltaic array maximum power point tracking techniques. *IEEE Transactions on Energy Conversion, 22*(2), 439–449.

Faranda, R., & Leva, S. (2008). Energy comparison of MPPT techniques for PV systems. *WSEAS Transactions on Power Systems, 3*(6), 446–455.

Femia, N., Petrone, G., Spagnuolo, G., & Vitelli, M. (2005). Optimization of perturb and observe maximum power point tracking method. *IEEE Transactions on Power Electronics, 20*(4), 963–973.

Fendri, D., & Chaabene, M. (2012). Dynamic model to follow the state of charge of a lead-acid battery connected to photovoltaic panel. *Energy Conversion and Management, 64*, 587–593.

Ghaisari, J., Habibi, M., & Bakhshai, A. R. (2007). An MPPT controller design for photovoltaic (PV) systems based on the optimal voltage factor tracking. *Proceedings of the IEEE conference on electrical power (EPC)* (pp. 359–362).

Gonzâlez-Longatt, F.M. (2005). Model of photovoltaic module in Matlab. *Proceedings of the second Iberoamerican conference of electrical, electronics and computation students* (pp. 1–5).

Guasch, D., & Silvestre, S. (2003). Dynamic battery model for photovoltaic applications. *Progress in Photovoltaics: Research and Applications, 11*(3), 193–206.

Hohm D.P., & Ropp M.E. (2000a). Comparative study of maximum power point tracking algorithms using an experimental, programmable, maximum power point tracking test bed. *IEEE proceedings of the photovoltaic specialists conference* (pp. 1699).

Hohm, D. P., & Ropp, M. E. (2000b). Comparative study of maximum power point tracking algorithms. *Progress in Photovoltaics: Research and Applications, 11*, 47–62.

Hussein, K. H., Muta, I., Hoshino, T., & Osakada, M. (1995). Maximum photovoltaic power tracking: An algorithm for rapidly changing atmospheric conditions. *Proceedings of the IEE Conference on Generation, Transmission and Distribution, 142*(1), 59–64.

Kenny, R.P., Friesen, G., Chianese, D., Bernasconi, A., & Dunlop, E.D. (2003). Energy rating of PV modules: Comparison of methods and approach. *Proceedings of the 3rd IEEE world conference on photovoltaic energy conversion* (pp. 2015–2018).

Khatib, T., Mohamed, A., & Sopian, K. B. (2012). A review of solar energy modeling techniques. *Renewable and Sustainable Energy Reviews, 16*(5), 2864–2869.

Kurella, A., & Suresh, R. (2013). Simulation of incremental conductance MPPT with direct control methods using cuck converter. *IJRET, 2*(9), 557–566.

Miveh, M. R., Rahmat, M. F., Ghadimi, A. A., & Mustafa, M. W. (2016). Control techniques for three-phase four-leg voltage source inverters in autonomous microgrids: A review. *Renewable and Sustainable Energy Reviews*, *54*, 1592–1610.

Moghadam, H., Tabrizi, F. F., & Sharak, A. Z. (2011). Optimization of solar flat collector inclination. *Desalination*, *265*(1), 107–111.

Oi, A. (2005a). *Design and simulation of photovoltaic water pumping system (Doctoral dissertation)*. San Luis Obispo, CA: California Polytechnic State University.

Oi, A. (2005b). Design and simulation of photovoltaic and IC water pumping system (Master course). San Luis Obispo, CA: University of Faculty of California Polytechnic State University.

Orgill, J. F., & Hollands, K. G. T. (1977). Correlation equation for hourly diffuse radiation on a horizontal surface. *Solar Energy*, *19*(4), 357–359.

Posadillo, R., & López Luque, R. (2009). Hourly distributions of the diffuse fraction of global solar irradiation in Córdoba (Spain). *Energy Conversion and Management*, *50*(2), 223–231.

Rana, G., Katerji, N., Lazzara, P., & Ferrara, R. M. (2012). Operational determination of daily actual evapotranspiration of irrigated tomato crops under Mediterranean conditions by one-step and two-step models: Multiannual and local evaluations. *Agricultural Water Management*, *115*, 285–296.

Reis, J.H., Miranda, A.C., Lemes, L.J., Viajante, G.P., & Chaves, E.N. (2015). Analysis of IMC method applied to output current control on DC/DC buck converter—a power led switching on application. *IEEE 13th Brazilian power electronics conference and 1st southern power electronics conference (COBEP/SPEC)* (pp. 1–6). IEEE.

Riggio, C., & Houghton, B. (2014). U.S. Patent No. 8,896,263. Washington, DC: U.S. Patent and Trademark Office.

Roger, J.A.; Perez, A.; Campana, D., Castiel, A.; & Dupuy, C.H.S. (1978). Calculations and in situ experimental data on a water pumping system directly connected to a 1/2 kW photovoltaic convertors array. *Proceedings of the photovoltaic solar energy conference* (pp. 1211–1220).

Salas, V., Olias, E., Barrado, A., & Lazaro, A. (2006). Review of the maximum power point tracking algorithms for stand-alone photovoltaic systems. *Solar Energy Materials and Solar Cells*, *90*(11), 1555–1578.

Salimi, M., Soltani, J., Markadeh, G. A., & Abjadi, N. R. (2013). Indirect output voltage regulation of DC-DC buck/boost converter operating in continuous and discontinuous conduction modes using adaptive backstepping approach. *Power Electronics, IET*, *6*(4), 732–741.

Sallem, S., Chaabene, M., & Kamoun, M. B. A. (2009a). Energy management algorithm for an optimum control of a photovoltaic water pumping system. *Applied Energy*, *86*(12), 2671–2680.

Sallem, S., Chaabene, M., & Kamoun, M. B. A. (2009b). Optimum energy management of a photovoltaic water pumping system. *Energy Conversion and Management*, *50*(11), 2728–2731.

Şen, Z. (2008). *Solar energy fundamentals and modeling techniques: Atmosphere, environment, climate change and renewable energy* (Vol. 276Springer, .

Soussi, M., Balghouthi, M., & Guizani, A. (2013). Energy performance analysis of a solar-cooled building in Tunisia: Passive strategies impact and improvement techniques. *Energy and Buildings*, *67*, 374–386.

Sweeeney, D. W., Graett, D. A., Bottcher, A. B., Lacario, S. J., & Camphll, K. L. (1987). Tomato yield and nitrogen recover as influenced by irrigation method, nitrogen source and mulches. *HortScience*, *22*, 27–29.

Taufik EE410 Power Electronics I—Lecture note. (2004). SanLuis Obispo, CA: Cal Poly State University.

Veerachary, M., & Yadaiah, N. (2000). ANN based peak power tracking for PV supplied DC motors. *Solar Energy*, *69*(4), 343–350.

Xiao, W., Dunford, W.G., & Capel, A. (2004, June). A novel modeling method for photovoltaic cells. *Proceedings of the 35th IEEE annual conference on power electronics specialists* (pp. 1950–1956).

Yahyaoui, I., Chaabene, M., & Tadeo, F. (2013). An algorithm for sizing photovoltaic pumping systems for tomato irrigation. *Proceedings of the IEEE conference on renewable energy research and applications (ICRERA)* (pp. 1089–1095).

Yahyaoui, I., Chaabene, M., & Tadeo, F. (2016). Evaluation of maximum power point tracking algorithm for off-grid photovoltaic pumping. *Sustainable Cities and Society, 25*, 65–73.
Yahyaoui, I., Sallem, S., Chaabene, M., & Tadeo, F. (2012). Vector control of an induction motor for photovoltaic pumping. *Proceedings of the international renewable energy conference (IREC)* (pp. 877–883).
Yin, X. X., Lin, Y. G., Li, W., Liu, H. W., & Gu, Y. J. (2015). Adaptive sliding mode back-stepping pitch angle control of a variable-displacement pump controlled pitch system for wind turbines. *ISA Transactions, 58*, 629–634.
Zhang, C. P., Sharkh, S. M., Li, X., Walsh, F. C., Zhang, C. N., & Jiang, J. C. (2011). The performance of a soluble lead-acid flow battery and its comparison to a static lead-acid battery. *Energy Conversion and Management, 52*(12), 3391–3398.

CHAPTER 3

Sizing Optimization of the Photovoltaic Irrigation Plant Components

3.1 INTRODUCTION

As it has previously been discussed in Chapter 2, Modeling of the Photovoltaic Irrigation Plant Components, system components modeling is a relevant step to understand the behavior of the PV water pumping system components, especially during the climatic parameters changes (Fig. 3.1). Since autonomous water pumping plant must be optimally sized to meet the criteria related to the plant autonomy, the water volume needed for irrigation, etc., in this chapter, the book reader can find a presentation and a validation of an approach for the PV system components sizing. Hence, a review on sizing algorithms in the literature is first detailed in Section 3.2. Then, the sizing algorithm is explained in Section 3.3. The sizing approach is tested and validated in a case study of the target area in Section 3.4.

Finally, the chapter conclusions are presented in Section 5.

3.2 A REVIEW ON SIZING ALGORITHMS IN THE LITERATURE

As the sizes of the PV installation components affect its autonomy (Kaldellis, Zafirakis, & Kondili, 2010; Sidrach-de-Cardona & Mora López, 1998), it is necessary to define some adequate values for the components parameters, such as the PVP surface and the number of batteries (Jakhrani, Othman, Henry Rigit, Samo, & Kamboh, 2012; Khatib, Mohamed, & Sopian, 2013).

Hence, researchers have established various methods to determine the optimum size of these components (Acakpovi, Xavier, & Awuah-Baffour, 2012). For instance, some works have focused on developing analytic methods based on a simple calculation of the panels surface and battery bank capacity using the energetic balance (Barra, Catalanotti, Fontana, & Lavorante, 1984; Groumpos & Papageorgiou, 1987; Shrestha & Goel, 1998). Other works have concentrated on the cost versus reliability question (Mellit, Benghanem, Hadj Arab, & Guessoum, 2003). Moreover, some researchers have proposed sizing algorithms based on the minimization of cost functions, using the loss of load probability (LLP) concept (Abouzahr & Ramakumar, 1991; Khatib, Mohamed, Sopian, & Mahmoud, 2012; Klein & Beckman, 1987; Maghraby, Shwehdi,

Specifications of Photovoltaic Pumping Systems in Agriculture
DOI: http://dx.doi.org/10.1016/B978-0-12-812039-2.00003-X

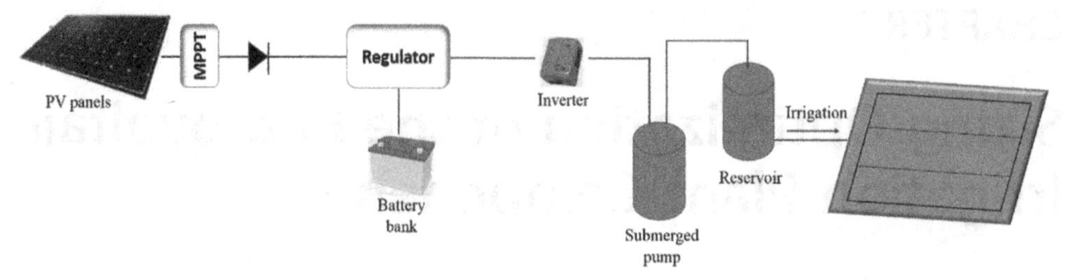

Figure 3.1 Proposed PV irrigation system.

& Al-Bassam, 2002; Yang, Zhou, Lu, & Fang, 2008). This LLP approach has also been combined with ANNs and genetic algorithms (Mellit et al., 2003; Yang et al., 2008). However, these methods may result in an oversized system for one location and an undersized one for another location (Mellit, Benghanem, & Kalogirou, 2007). The oversized case results in high installation costs. With an undersized case, the installation is unable to supply the load with the energy needed. Moreover, the installation lifetime is shorter, due to the excessive use of batteries. For these reasons, the sizes must be carefully selected for each specific application and location (Mellit et al., 2007).

In this context, several tools for PV installation sizing are available. For instance, HOMER (www.homerenergy.com), COMPASS (www.lorentz1.software.informer), PVsyst (www.pvsyst.com), RAPSim (www.sourceforge.net), and RETScreen International (www.retscreen.net) determine the optimum size of the PV component. Table 3.1 summarizes the references for these tools and gives comments on their design methodologies.

The tools presented in Table 3.1 determine the optimum size of the PV installation components by taking into account the energetic, economic, and environmental aspects (Ben Ammar, 2011). However, some softwares (such as COMPASS) do not include batteries. Hence, in the case of water pumping, it is limited to the case of pumping over the sun. Additionally, HOMER is a good tool for sizing. Despite it guarantees the installation autonomy, it may give an oversized sizing, since it concentrates in the system autonomy. PVsyst is a good tool for sizing since it takes into account the load loss probability. However, the evaluation of the sizing efficiency can be done by years. RAPSim focuses on modeling alternative power supply options. Using costs calculation throughout the lifespan, this tool is very adequate to predict the system performance and economic parameters of hybrid PV–Wind–Diesel–Battery systems (Bernal-Agustín & Dufo-López, 2009). Moreover, RETScreen is an excel tool that assists the user in determining the energy production, life cycle costs, and greenhouse gas emission reductions for various types of renewable energy (Bernal-Agustín & Dufo-López, 2009). This tool allows the electricity produced to be determined using statistical sizing (provided by the user), models, and climatic data of the target site (Ben Ammar, 2011).

Table 3.1 Summary of sizing software (Ben Ammar, 2011)

Software	Organism	Observations
HOMER (www.homerenergy.com)	NREL: National Renewable Energy Laboratory, USA	Components classified by the cost and life cycle
COMPASS (www.lorentz1.software.informer)	Global Headquarters and Technology Center, Hamburg, Germany	Batteries are not included in the software library
PVsyst (www.pvsyst.com)	University of Geneva, Switzerland	Sizing efficiency is done by years
RAPSim (www.sourceforge.net)	Murdoch University Energy Research Institute, Australia	Sizing is based on the evaluation of the installation yield using different components configurations
RETScreen International (www.retscreen.net)	The Ministry of Natural Resources Canada	The conception is based on statistical models to evaluate the economic and energetic balance

Hence, these tools may give a good sizing for the installation autonomy, but they may result in oversized components. In this context, based on the load demand and the climatic data, an algorithm to determine the optimum size of the PV installation components (Fig. 3.1), using models of the PVPs, the battery bank and the pump, and measured climatic data, is proposed.

3.3 SIZING ALGORITHM PROPOSAL

A good sizing must fulfill that the installation provides the electrical demand of the load (Yahyaoui, Chaabene, & Tadeo, 2013). Hence, during the months that correspond to the vegetative cycle of the crops, knowing the water volume needed for irrigating the crops, the site characteristics, the solar radiation, and the PVP type, the values selected of the PV panels surface, the battery bank capacity, and the reservoir volume must guarantee the water volume needed for the crops irrigation, the system autonomy, and the battery bank safe operating (Khatib et al., 2013). Indeed, the idea consists in calculating the values that guarantee, on the one hand, the balance between the charged and discharged energy in the battery bank, and on the other hand, the pumping of the water volume needed. It is important to point out that the components size chosen must fulfill the irrigation requirements for all the months of the crops' vegetative cycle (*March* to *July*). Therefore, the algorithm's main objective is to ensure the load supply throughout the day, while protecting the battery against deep discharge or excessive charge and guaranteeing the water volume needed for the irrigation. The scheme of the proposed approach is presented in Fig. 3.2 (Yahyaoui et al., 2013). The algorithm depends on:

- the water volume needed,
- the site characteristics,
- the panel characteristics.

The proposed sizing approach aims to find the optimum panels' surface S_{opt} and the batteries number $n_{bat_{opt}}$ that guarantee the installation autonomy when supplying the pump and pumping the water volume needed for the crops irrigation. Hence, the idea consists in searching the optimal components sizes that ensure the balance between the charged and the extracted energies E_c and E_e, respectively (Fig. 3.3). In fact, the battery bank supplies the load when the panel does not generate the

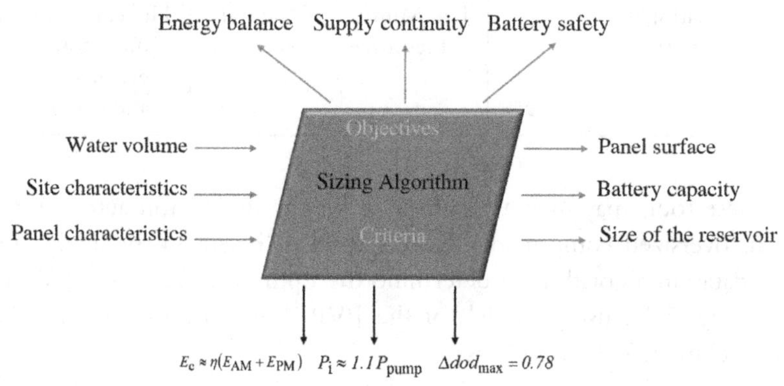

Figure 3.2 Planning of the proposed sizing algorithm.

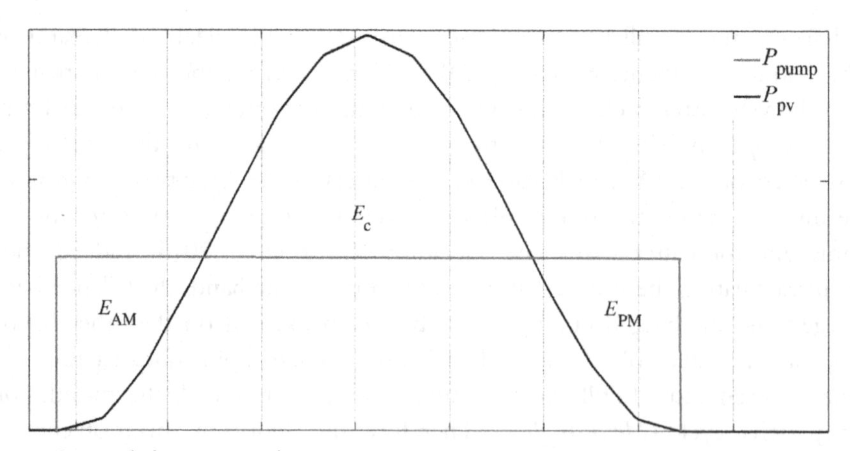

Figure 3.3 Energy balance principle.

sufficient power and is charged with the PV energy produced in excess. The energy balance can be expressed as follows:

$$E_c \approx E_{AM} + E_{PM} \quad (3.1)$$

The sizing algorithm is performed using two sub algorithms during the crops' vegetative cycle (*March* to *July*): The first Algorithm 3.1 allows the size of the panel surface S_M and the number of batteries n_{bat_M} to be determined for each month M. Then, Algorithm 3.2 is performed to deduce the final system components sizes.

For the application studied in this book, the PVP yield-based model is used for the sizing algorithm. In fact, since the proposed sizing algorithm uses only the power curve of the PV modules, a yield-based model has been selected. Algorithm 3.1 is detailed now in steps following the approach presented in Fig. 3.4.

3.3.1 Algorithm 3.1: determination of S_M and n_{bat_M}

The algorithm 3.1 is described by the following steps:

Step 1. Estimation of the diffused and direct radiation using Eqs. (2.7) and (2.13) (Posadillo & López Luque, 2009).

Step 2. Deduction of the solar radiation $H_t(t, d)$ in a tilted panel using Eq. (2.14) (El-Sebaii, Al-Hazmi, Al-Ghamdi, & Yaghmour, 2010; Posadillo & López Luque, 2009).

Step 3. Estimation of the cell temperature $T_c(t)$ using Eq. (2.21) (Arunkumar et al., 2012; Clean Energy Decision Support Centre, 2001–2004; Collares-Pereira & Rabl, 1979; Erbs, Klein, & Duffie, 1982; Moghadam, Tabrizi, & Sharak, 2011; Orgill & Hollands, 1977; Şen, 2008).

Step 4. Deduction of the panel yield $\eta_{pv}(t)$ using Eq. (2.20) (Chaabene, 2009; Yahyaoui et al., 2013).

Step 5. Calculation of the crops water needs V: The determination of the water volume needed for tomato growth is essential to define the amount of water to be pumped. The water volume depends essentially on the crop growth stage and the evapotranspiration (Rana, Katerji, Lazzara, & Ferrara, 2012). In the literature, many models have been used to describe the evapotranspiration. For instance, some researchers used the Penman method, which depends essentially on the net radiation at the crop surface, the mean air temperature, and the wind speed (Gong, Xu, Chen, Halldin, & Chen, 2006). Other works presented some models to describe the evapotranspiration, such as the Thornthwaite method, which depends on the sunlight duration and the air temperature (Obid, Khaleel, & Nife, 2013). The Blaney–Criddle method has also been used (Kashyap & Panda, 2001; Obid et al., 2013). This method includes the seasonal crop coefficient k_c, in addition to the sunlight duration and the air temperature, which provides better patterns of the

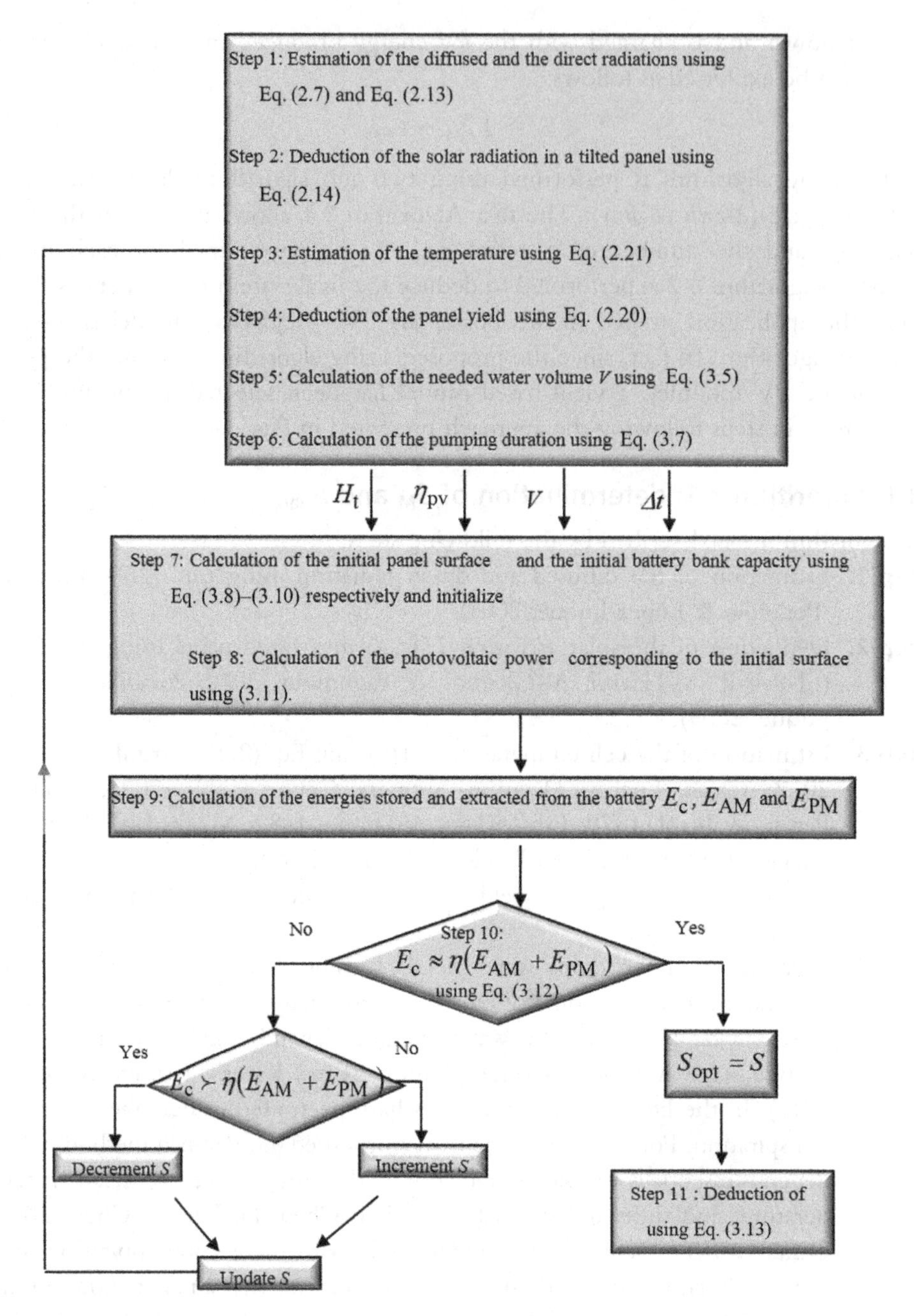

Figure 3.4 Sizing Algorithm 3.1 for each month M.

needed water volume. For this reason, the Blaney–Criddle method is used in this study.

The daily water volume V_n, required by the crop is given by (Rana et al., 2012):

$$V_n = k_c\, ET_0 \quad (3.2)$$

where:

k_c: the monthly crop growth coefficient,

ET_0: the monthly reference evapotranspiration average, which depends on the ratio of the mean daily daytime hours for a given month to the total daytime hours in the year p and the mean monthly air temperature T for the corresponding month, is evaluated (Kashyap & Panda, 2001; Obid et al., 2013, Yahyaoui et al., 2016):

$$ET_0 = K\, p(0.46\, T + 8.13) \quad (3.3)$$

where K is the correction factor, expressed by (Obid et al., 2013):

$$K = 0.03\, T + 0.24 \quad (3.4)$$

To obtain the necessary gross water, it is essential to estimate the irrigation losses. For this, an additional water quantity must be provided for the irrigation to compensate for those losses. Thus, the final recommended water volume is evaluated as follows (Beltrán, 1999):

$$V = (k_c ET_0 - r_m)\left(1 + \frac{1 - l_f(1 - L_R)}{l_f(1 - L_R)}\right) \quad (3.5)$$

where:

r_m: the average monthly rain volume,

l_f: leaching efficiency coefficient as a function of the irrigation water applied (Van Hoorn, 1981),

L_R: the leaching fraction given by the humidity that remains in the soil, expressed by (Letey et al., 2011):

$$L_R = \frac{EC_w}{5\, EC_e - EC_w} \quad (3.6)$$

EC_w: the electrical conductivity of the irrigation water (dS/m),

EC_e: the crop salt tolerance (dS/m).

Step 6. Calculation of the pumping duration Δt, using the water flow Q. In this application, the pump flux is constant. Thus, Δt can be evaluated as follows:

$$\Delta t = \frac{P_{pump}}{Q} \quad (3.7)$$

Step 7. Calculation of the minimum panel surface S_i and the initial battery number n_{bat_i} using Eqs. (3.8–3.10) respectively, based on the irrigation frequency (Yahyaoui et al., 2013):

$$S_i = \frac{P_{pump}\Delta t}{W_{pv}\,\eta_{bat}^{2}\,\eta_l\,\eta_{pv}\,\eta_{reg}\,\eta_{inv}\,\eta_{opt\ ther}\,\eta_{matching}}\left(1+\frac{d_{aut}}{d_{rech}}\right) \tag{3.8}$$

$$E_c = E_{tot}\,\Delta dod_{max} = E_d\,d_{aut} \tag{3.9}$$

Hence:

$$n_{bat_i} = \frac{E_d\,d_{aut}}{V_{bat}\,C_{bat}\Delta dod_{max}} \tag{3.10}$$

where:
P_{pump}: the pump power (W),
Δt: the water pumping duration (h),
d_{aut}: the days of autonomy,
d_{rech}: the days needed to recharge the battery,
W_{pv}: the average daily radiation ($Wh/m^2/day$),
η_{bat}: the electrical efficiency of the battery bank,
η_l: the electrical efficiency of the installation that includes the ohmic wiring and mismatching wiring losses,
η_{pv}: the efficiency of each PVP,
η_{reg}: the regulator performance,
η_{inv}: the inverter performance,
$\eta_{opt\ ther}$: the panel performance facing to optical and thermal effects (%),
$\eta_{matching}$: the panel matching performance (%),
E_d: the daily consumption (Wh),
V_{bat}: the battery voltage (V),
Δdod_{max}: the maximum *dod* variation (%),
C_{bat}: the nominal capacity for one battery (Ah).

Step 8. Calculation of P_{pvi} corresponding to the minimum panel surface S_i, using the following Eq. (3.11) (Yahyaoui et al., 2013):

$$P_{pv\ i} = \eta_{pv}\,\eta_{opt\ ther}\,\eta_{reg}\,\eta_{matching}\,S_i\,H_t \tag{3.11}$$

where:
η_{pv}: the panels yield (%) (Eq. (2.20)),
$\eta_{opt\ ther}$: the panel performance facing to optical and thermal effects (%),
$\eta_{matching}$: the panel matching performance (%),
H_t: the solar radiation on a tilted panel (W/m^2) (Eq. (2.14)),
S_i: the initial panel surface (m^2).

Step 9. Calculation of the energies expected to be stored and extracted from the battery each day by evaluating the area E_c and E_e, respectively (Fig. 3.3).

Step 10. If the discharged energy is higher than the charged energy, the algorithm increases the panel surface by the minimum increment of the PVP size commercially available: The algorithm looks for the best configuration to guarantee the balance between the demanded and the produced energies, by ensuring the equality between the charged E_c and discharged energies E_e in the battery bank (Eq. (3.1)).

The balance between the charged and the extracted energies E_c and E_e, respectively, does not guarantee the system autonomy, due to the fluctuation in the solar radiation and the energy losses in the installation components. Thus, to ensure the system autonomy and protect the battery against deep discharges, the algorithm is performed by adopting an efficiency coefficient η that allows the $\Delta\, dod$ to be less than $\Delta\, dod_{max}$ (η is equal to $1.28 * \eta_{error}$ in this case, where η_{error} describes the error between the clear sky and measured solar energies). Thus, Eq. (3.1) becomes:

$$E_c \approx \eta(E_{AM} + E_{PM}) \tag{3.12}$$

Moreover, to ensure the continuity of the load supply, the previous condition is performed with $P_i = 1.1\ P_{pump}$.

Step 11. Deduction of n_{bat_M} (Eftichios, Kolokotsa, Potirakis, & Kalaitzakis, 2006; Yahyaoui et al., 2013; Zhang et al., 2011):

$$n_{bat_M} = \frac{E_c}{C_{bat}{}^{k_p}} \tag{3.13}$$

where:

E_c: the energy charged in the battery bank (Wh),

C_{bat}: the nominal capacity for one battery (Ah).

3.3.2 Algorithm 3.2: deduction of S_{opt} and C_{opt}

Using Algorithm 3.2, presented in Fig. 3.5, the final values of the panel surface S_{opt} and the capacities number $n_{bat_{opt}}$ are then deduced. S_{opt} corresponds to the maximum value of the panel surface obtained during the months. The optimum batteries number is the corresponding value for S_{opt}, since it is the most critical month.

3.4 APPLICATION TO A CASE STUDY

The sizing Algorithm 3.2 is applied now to evaluate the components sizes of the studied installation. The proposed algorithm is tested during the months that correspond to the vegetative cycle of tomatoes (*March* to *July*), using data of the target area. The sizing

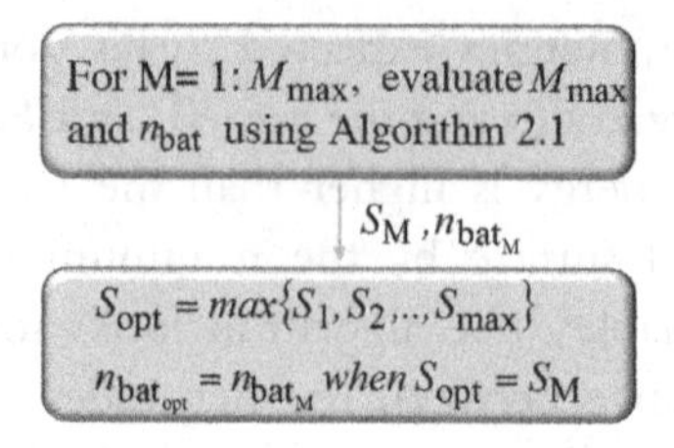

Figure 3.5 Sizing Algorithm 3.2.

Table 3.2 Components parameters

Parameters	Values
η_{bat}	90% (Yahyaoui et al., 2013)
η_{inv}	92% (Chaabene, 2009)
η_l	95% (Yahyaoui et al., 2013)
$\eta_{matching}$	80% (Chaabene, 2009)
$\eta_{opt\ therm}$	90% (Chaabene, 2009)
η_{reg}	90% (Yahyaoui et al., 2013)
η_r	10.58% (Chaabene, 2009)

results are validated using measured data of the target area and later using HOMER and PVsyst.

3.4.1 Sizing for the case study

To carry out the sizing in the case study of the 10 ha land surface of Medjez el Beb (Northern of Tunisia), the proposed Algorithms 3.1 and 3.2 are tested using the parameters presented in Table 3.2, together with the measured climatic data of the target area.

To evaluate the solar radiation on a tilted panel in the target area (Medjez El Beb, Tunisia), the tilted angle has been chosen based on PVsyst analysis (Figs. 3.6 and 3.7). In fact, considering the latitude angle as the tilted angle allows the transposition factor TF to be near to its optimum value ($TF_{annual} = 96\%$ for the annual and $TF_{monthly} = 90\%$ for the sunny months).

In the sizing approach, the TE500CR panel is considered. The solar radiation G is presented in Fig. 3.8, the ambient temperature T_a in Fig. 3.9, and the average monthly rain volume r_m, the monthly reference evapotranspiration ET_0, and the monthly crop growth coefficient k_c in Fig. 3.10.

The Algorithm 3.1 was first evaluated. That is, the solar radiation accumulated on a tilted panel is evaluated using Eq. (2.14). Then, the panel yield is calculated for each month using Eq. (2.20) (Table 3.3). In parallel, the water needed V is evaluated

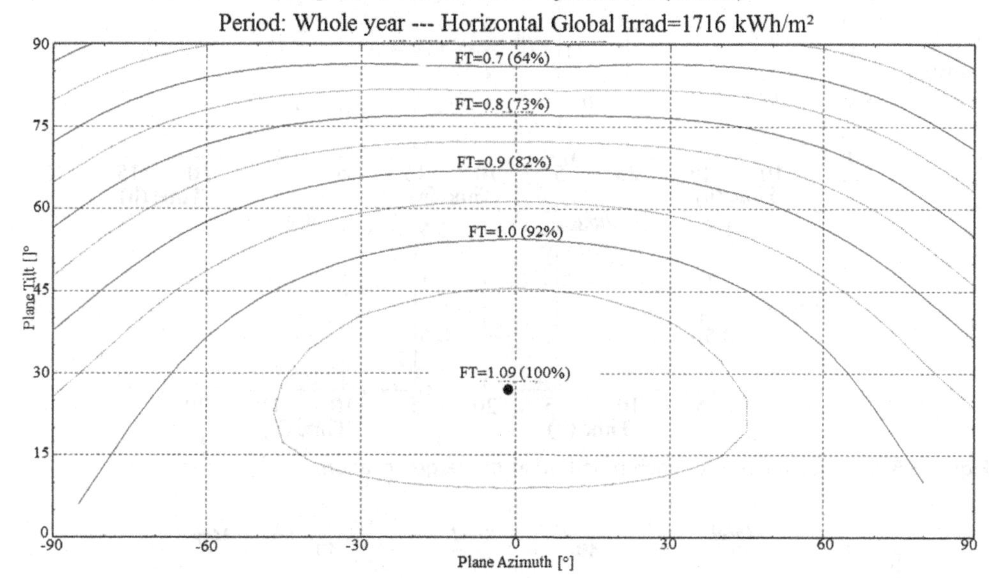

Figure 3.6 Annual transposition factor for Medjez El Beb.

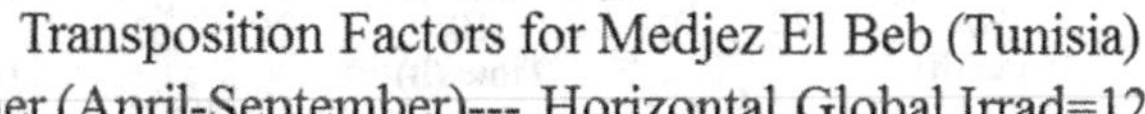

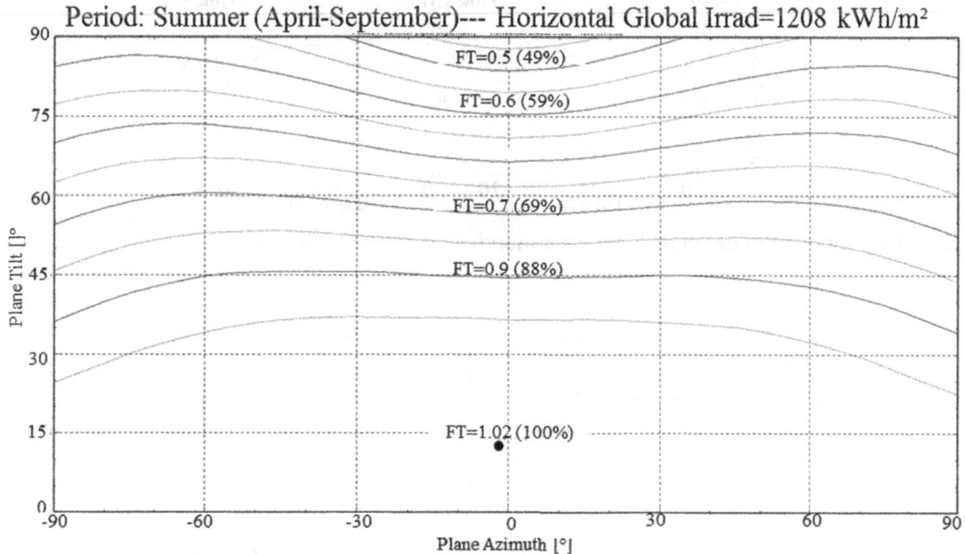

Figure 3.7 Monthly transposition factor for Medjez El Beb.

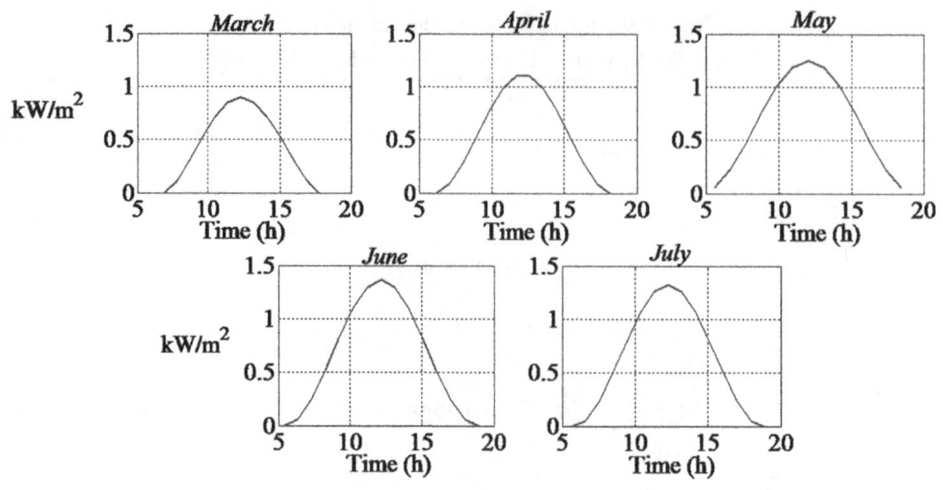

Figure 3.8 Solar radiation G for each month M at the target location.

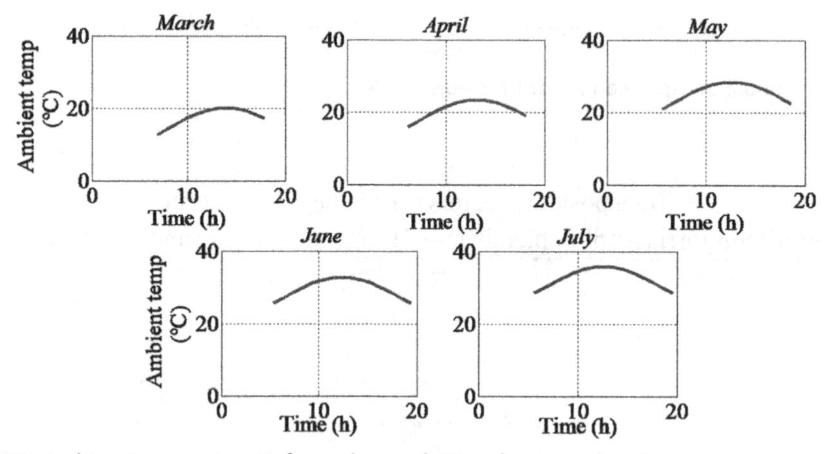

Figure 3.9 Ambient temperature T_a for each month M at the target location.

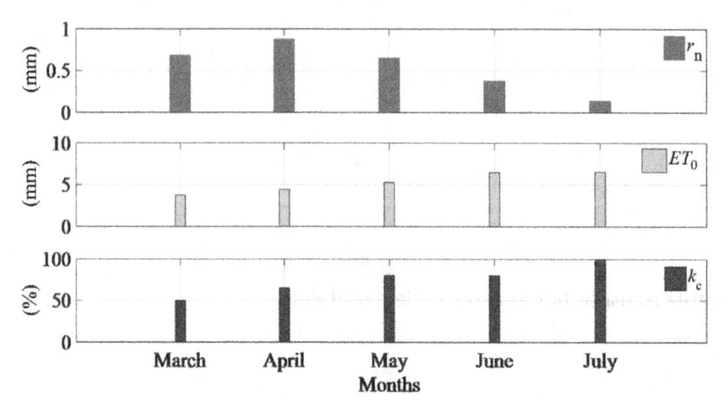

Figure 3.10 Crops characteristics and daily rain estimation for each month M.

Table 3.3 Climatic parameters, panel efficiency, and irrigation parameters estimated for the case study

Months	***March***	***April***	***May***	***June***	***July***
Results					
T_a (°C)	14	17.25	20	22	30
$\overline{H}$ (Wh)	4023.6	5512.3	5815.2	7392.2	7163.2
k_t (%)	54	51	54	61	64
W_{pv} (Wh) (Eq. (2.18))	5908.6	7562.1	8030.9	9479.0	9136.7
η_{pv} (%) (Eq. (2.20))	10.16	10.06	9.91	9.75	9.37
Water volume m^3/10 ha (Eq. (3.5))	60.70	100.37	179.82	241.10	321.03
Pumping duration Δt (h) (Eq. (3.7))	2.5	4.13	7.41	9.93	13.25

Table 3.4 Initial values of the panels surface and number of batteries

Months	***March***			***April***			***May***
Results							
W_{pvc} (clear sky model) (Wh)	5760			7180			8120
Maximum number of cloudy days per month n_c (Eq. (3.14))	9			7			9
Clouds rate per day A_c (%) (Eq. (3.15))	30.15			23.23			28.38
Irrigation frequency f_i (Chapter: Renewable Energies and Irrigation)	3	3	2	2	2	1	1
d_{aut}	1	1	1	1	1	1	1
d_{rech}	3	3	2	2	2	1	1
Initial panel surface (m^2) (Eqs. (3.8) and (3.9))	61	61	68	89	89	107	203
Initial numbers of batteries n_{bat_i} (Eq. (3.10))	4	4	4	5	5	5	10

Months	***June***	***July***		
Results				
W_{pvc} (clear sky model) (Wh)	8500	8340		
Maximum number of cloudy days per month n_c (Eq. (3.14))	3	4		
Clouds rate per day (%) A_c (Eq. (3.15))	13.03	14.11		
Irrigation frequency f_i (Chapter: Renewable Energies and Irrigation)	1	1	2	2
d_{aut}	1	1	1	1
d_{rech}	1	1	2	2
Initial panel surface (m^2) (Eqs. (2.61) and (3.8))	234.5	337	168.5	168.5
Initial numbers of batteries n_{bat_i} (Eq. (3.10))	14	18	18	18

depending on the crops vegetative cycle and the site characteristics using Eq. (3.5), and thus, the pumping duration Δt is deduced using Eq. (3.7).

The initial values of S_i and n_{bat_i}, summarized in Table 3.4, are used to test the condition Eq. (3.1). Indeed, if the charged energy is higher than the discharged energy, the panel surface is increased by the minimum panel available surface in the market (in this case, the increment is 0.5 m^2), and vice versa.

The maximum number of cloudy days per month n_{c_i} and the amount of clouds per day A_{c_i} are evaluated for each month M to deduce the days of autonomy. They are calculated using Eqs. (3.14) and (3.15):

$$n_{c_i} = \frac{n_{M_i} \cdot (W_{pvc_i} - \overline{H}_{t_i})}{(1 - DA_i) W_{pvc_i}} \tag{3.14}$$

$$A_{c_i} = \frac{W_{pvc_i} - \overline{H}_{t_i}}{W_{pvc_i}} \tag{3.15}$$

where:

n_{M_i}: the days number in the month M,

W_{pvc_i}: the solar energy for the month M using the clear sky model (Wh),

$\overline{H}_{t_i}$: the solar energy for the month M using the clear database (Wh),

DA_i: the ratio between diffuse and global daily solar radiation.

Algorithm 3.1 results are summarized in Table 3.5 and evaluated in Figs. 3.11 and 3.12. They show that the proposed strategy always ensures the needed water volume, respects the battery bank depth of discharge limits and the energy balance. In fact, Figs. 3.11 and 3.12 prove that the proposed algorithm guarantees the needed water volume for the crops irrigation, since the pump is supplied by the PV panels and the battery bank. This has been proved for all the months of the crops vegetative cycle,

Table 3.5 Summary of the results of the calculation of the minimum panel surface and number of batteries needed each month *M*.

Months	***March***	***April***	***May***	***June***	***July***
Results					
η_{error}	1.30	1.23	1.28	1.13	1.14
$E_{AM} + E_{PM}$ (Wh/day)	10,991	14,481	10,239	12,511	24,046
E_c (Wh/day)	18,725	23,035	16,807	18,033	35,314
E_{pump} (Wh/day)	11,258	18,615	33,350	44,716	59,541
E_{PV} (Wh)	20,371	29,296	43,378	55,035	82,802
S_M (m^2)	**37.5**	**41.5**	**54.5**	**61.5**	**101.5**
n_{bat_M} (Eq. (3.13))	**4**	**5**	**4**	**5**	**8**
η (Eq. (3.12))	***1.66***	***1.57***	***1.64***	***1.44***	***1.46***
$\eta_1 = \frac{E_c}{E_{AM} + E_{PM}}$ (Eq. (3.12))	***1.7***	***1.59***	***1.64***	***1.44***	***1.47***

η_{error} : Error between the clear sky and measured solar energies; $E_{AM} + E_{PM}$: Energy extracted from the battery bank; E_c : Energy saved in the battery bank; E_{pump}: Energy supplied to the pump; E_{PV}: PV energy; S_M: PV panels surface; n_{batM} : Batteries number; n: theorotical quotient between the saved and the extracted energies of the battery bank; n_1 : Quotient between the saved and the extracted energies of the battery bank obtained by the algorith; The bold numbers are the algorithm results.

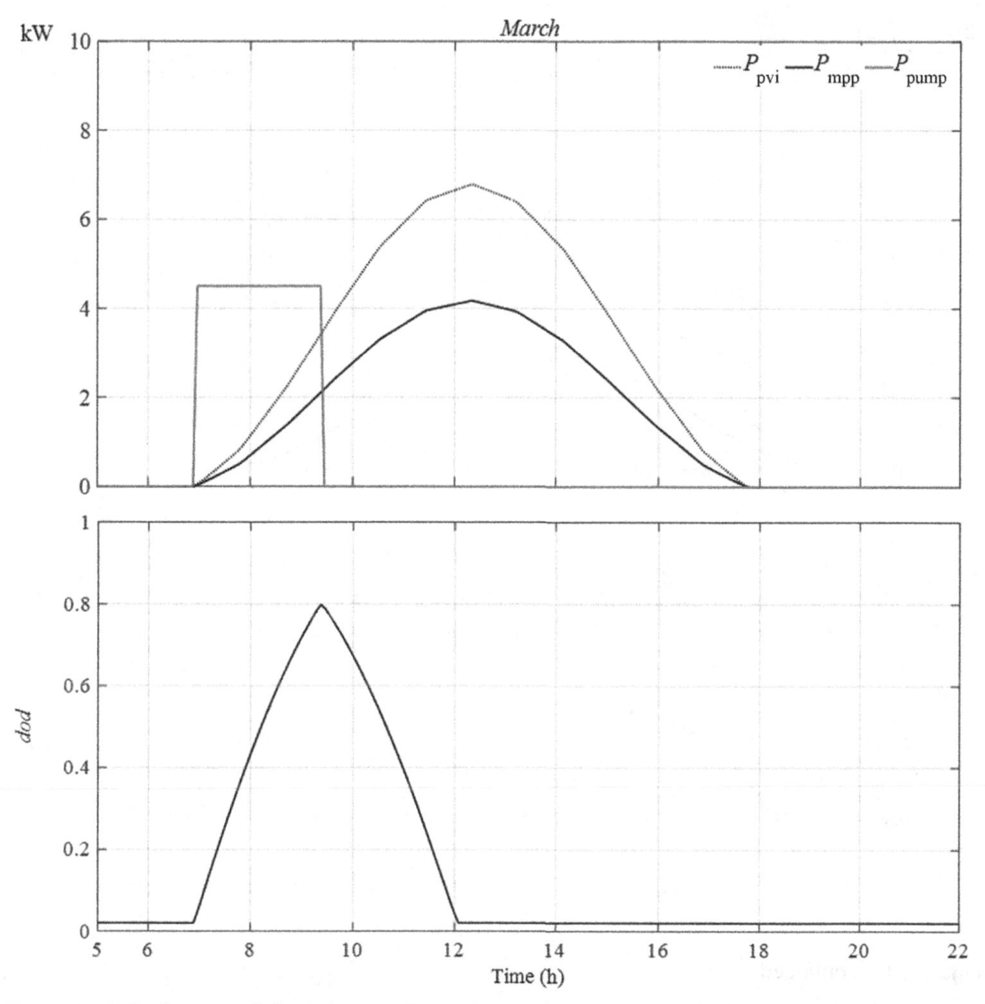

Figure 3.11 Evaluation of Algorithm 3.1 for each month using mean climatic data values.

using the daily clouds amount. Moreover, this algorithm ensures the energy balance for each month M: In Table 3.5, the efficiency coefficient η is around the fixed values throughout all the considered months. For this value, $\Delta\, dod_{max}$ is guaranteed to be equal to 0.78 ($\Delta\, dod_{max} = dod_{max} - dod_{min}$). For instance, in *July*, η is fixed to be equal to 1.46. The obtained value η_1 with the algorithm is equal to 1.47. Moreover, in *March*, the generated PV power during the morning is used to supply the pump together with the battery bank during the pumping duration. Then, the PV power generated is used to charge the battery bank for the rest of the day hours. The quotient between the cumulated and extracted energies is equal to 1.66, which is near to the value initially fixed in Algorithm 3.1 ($\eta = 1.7$). Thus, the extracted energy (E_e) is almost equal to the accumulated energy (E_c).

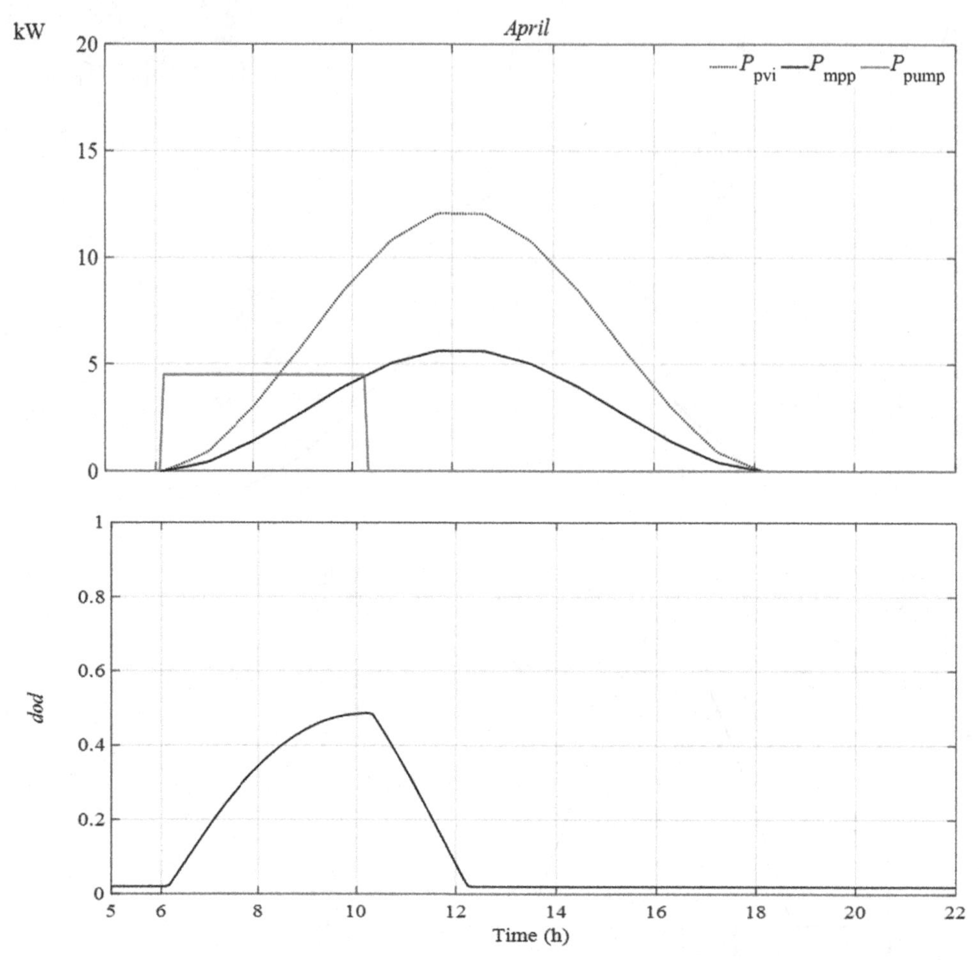

Figure 3.11 Continued.

For the energy balance, an error coefficient obtained by the evaluation of the daily clouds amount is used. Hence, in this study, the possibility of having cloudy days are taken into account. For example, in *April* the amount of clouds per day is 23.23%. Figs. 3.11 and 3.12 prove that the obtained panels surface and battery bank capacity obtained by Algorithm 3.1 satisfy the energy balance. In other terms, all the stored energy is consumed. This is possible thanks to the calculation of the number of batteries, which is done by considering the same $\Delta\, dod_{max}$ value that can be reached, for all the months ($\Delta\, dod_{max} = 0.78$). Hence, the obtained surface allows the load to be supplied during the requested pumping duration Δt and also provides the energy E_c needed to charge the battery bank.

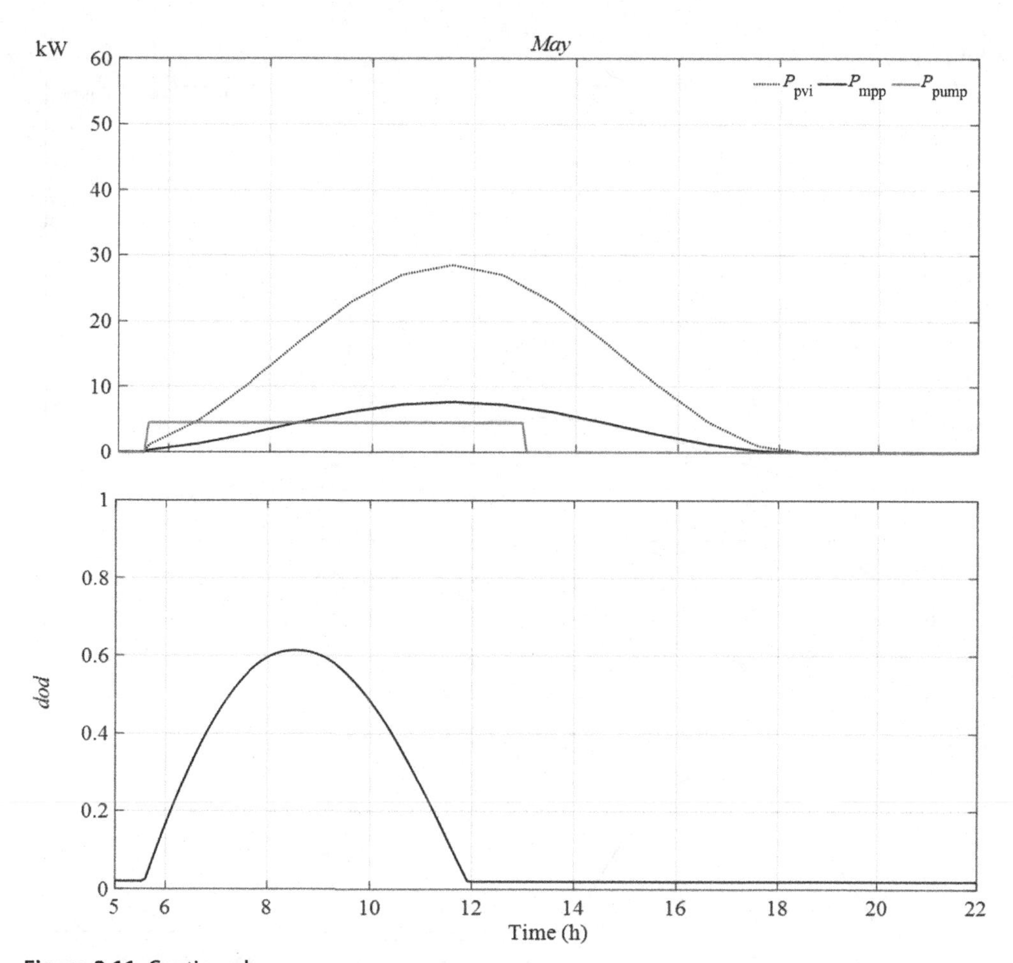

Figure 3.11 Continued.

Using Algorithm 3.2, the final size of the panel surface and the batteries number are deduced. Hence, $S_{opt} = 101.5\ m^2$, $n_{bat_{opt}} = 8$ batteries (210 Ah/12 V). These values correspond to the results of the Algorithm 3.1 for *July*, since it is the most critical month for the crops irrigation. This result will be used, in the following subsection, to evaluate the sizing algorithm results using measured climatic data.

3.4.2 Validation using measured climatic data

To demonstrate the efficiency of the sizing algorithm, measured data of the solar radiation G and ambient temperature T_a of the target area are used, for the months of tomatoes' vegetative cycle (Figs. 3.13–3.14).

The charged and extracted energies E_c and E_e have been evaluated using Algorithm 3.1. The obtained values of η_{exp} for *July*, presented in Table 3.6, prove that

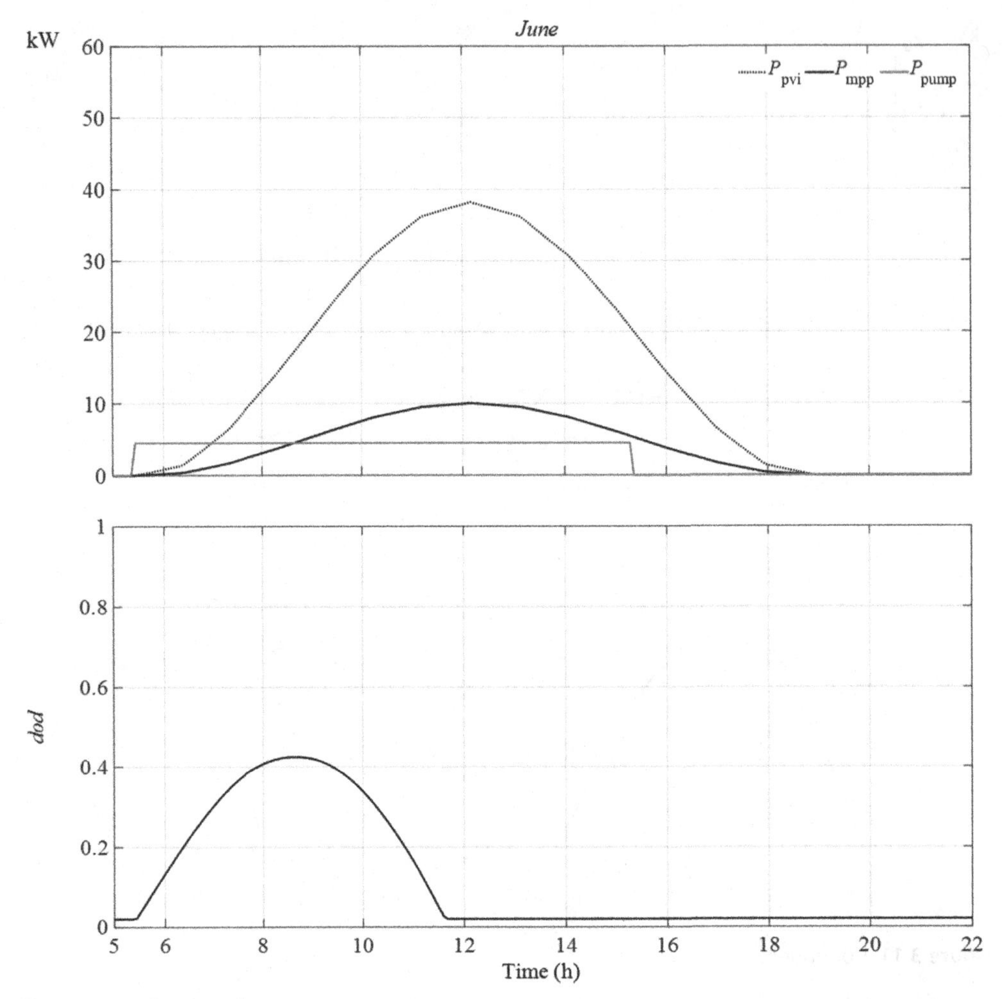

Figure 3.11 Continued.

the approach is validated with experimental data from G and T_a, since the values of η_{exp} are close to η and hence the pump is supplied during the pumping period (Figs. 3.15 and 3.16). For the rest of the months, the charged energy E_c is higher than the energy in demand, since the components size that corresponds to *July* is used (which is the most critical month) (Ben Salah & Ouali, 2012).

3.4.3 Validation using HOMER

In order to test *the sizing* approach, the installation sizes obtained by the algorithm proposed are now compared with the components sizes obtained using HOMER. In fact, using this tool, the needed panels surface is 142 m^2, and the battery number is

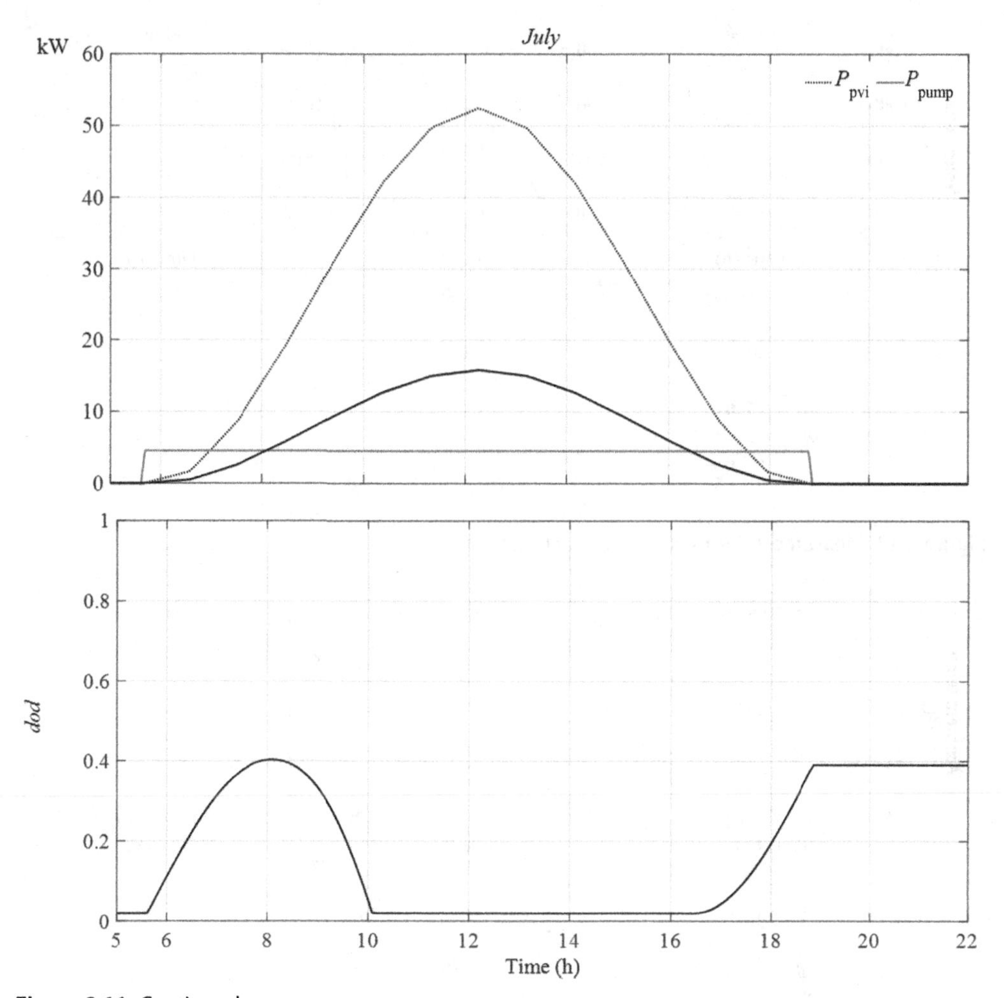

Figure 3.11 Continued.

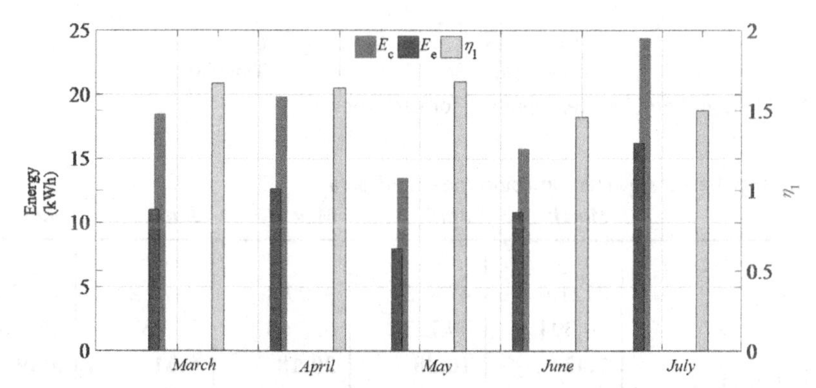

Figure 3.12 Summary of the daily energies using mean climatic data values for each month *M* using Algorithm 3.1.

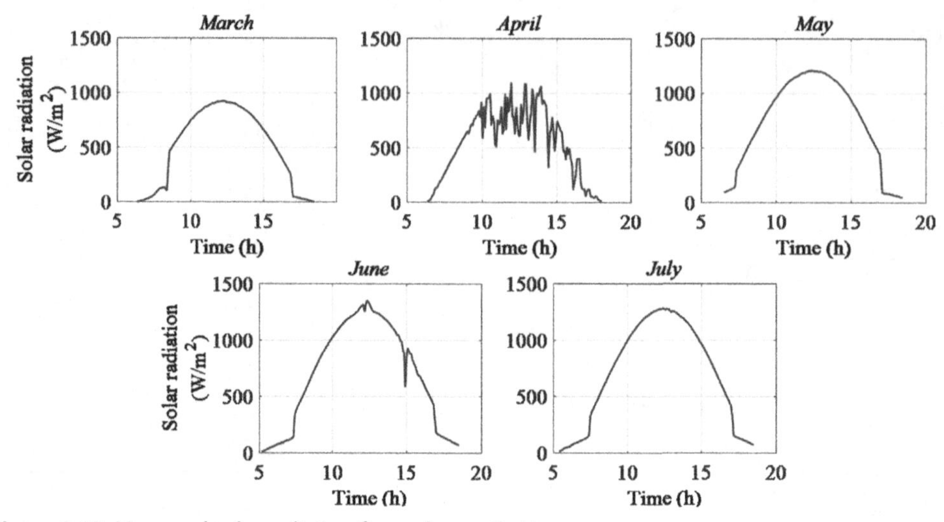

Figure 3.13 Measured solar radiation for each month *M*.

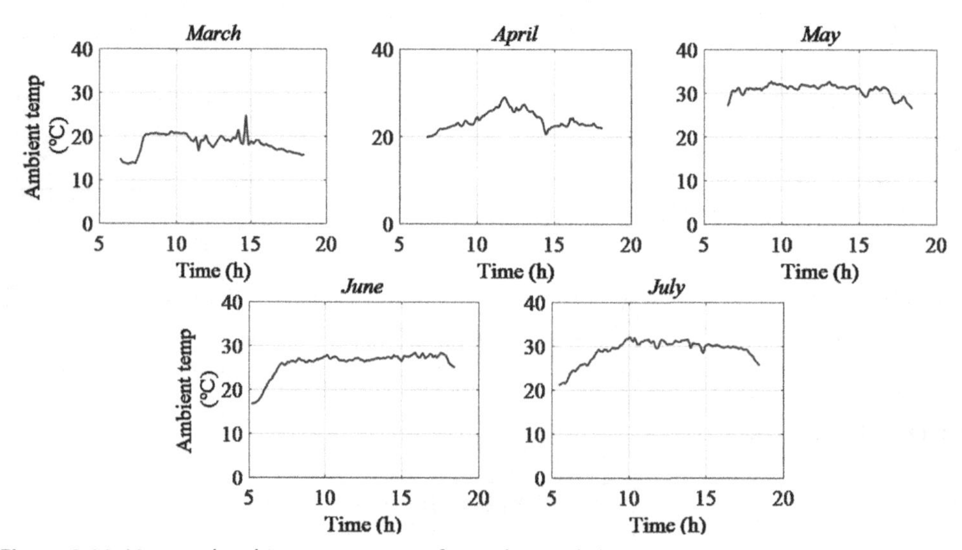

Figure 3.14 Measured ambient temperature for each month *M*.

Table 3.6 Energy balance evaluation using measured data

Months	*March*	*April*	*May*	*June*	*July*
Results					
$E_{AM} + E_{PM}$ (Wh)	8121.6	4740.5	3109.4	8572.2	12,855.0
E_c (Wh)	60,394	49,727	62,763	56,737	46,269
$\eta_{exp} = \frac{E_c}{E_e}$ (Eq. (3.12))	**7.43**	**10.48**	**20.18**	**6.61**	**3.59**

Bold values denotes Quotient between the saved and the extracted energies of the battery bank using experimental climatic data.

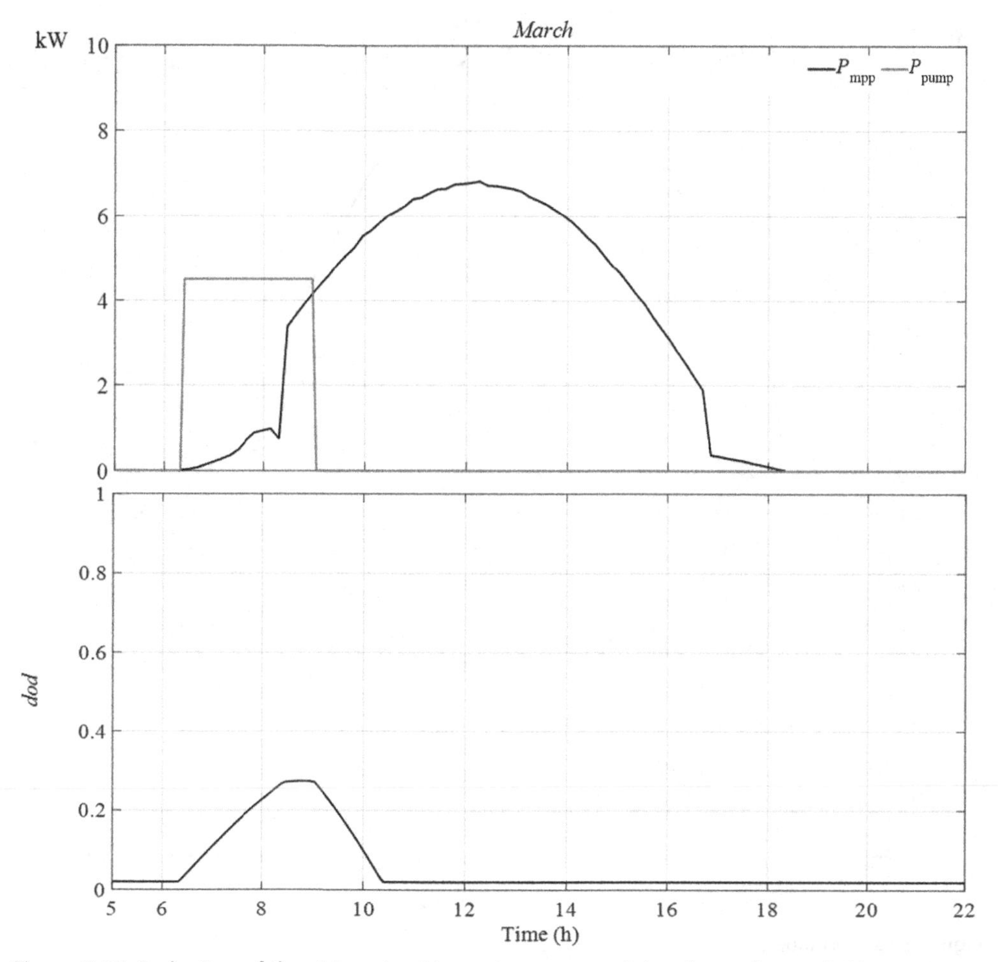

Figure 3.15 Evaluation of the sizing algorithm using measured data for each month *M*.

14 batteries (210 Ah/12 V). The simulation results, presented in Figs. 3.17 and 3.18, show that the panels size suggested by HOMER is higher than the surface obtained by the studied algorithm, since HOMER considers the cases when the battery bank do not operate, and allows big importance to the day autonomy. Hence, the installation components are oversized by HOMER.

3.4.4 Validation using PVsyst

The installation size is also tested using PVsyst. The simulation shows that the adopted size ($S = 101.5\ m^2$ and $n_{bat} = 8$ batteries/210 Ah/12 V) gives good results (Fig. 3.19). In fact, during the crops vegetative cycle, the solar fraction (which is defined by the quotient of the available by the needed powers of the load) is equal to one, except in

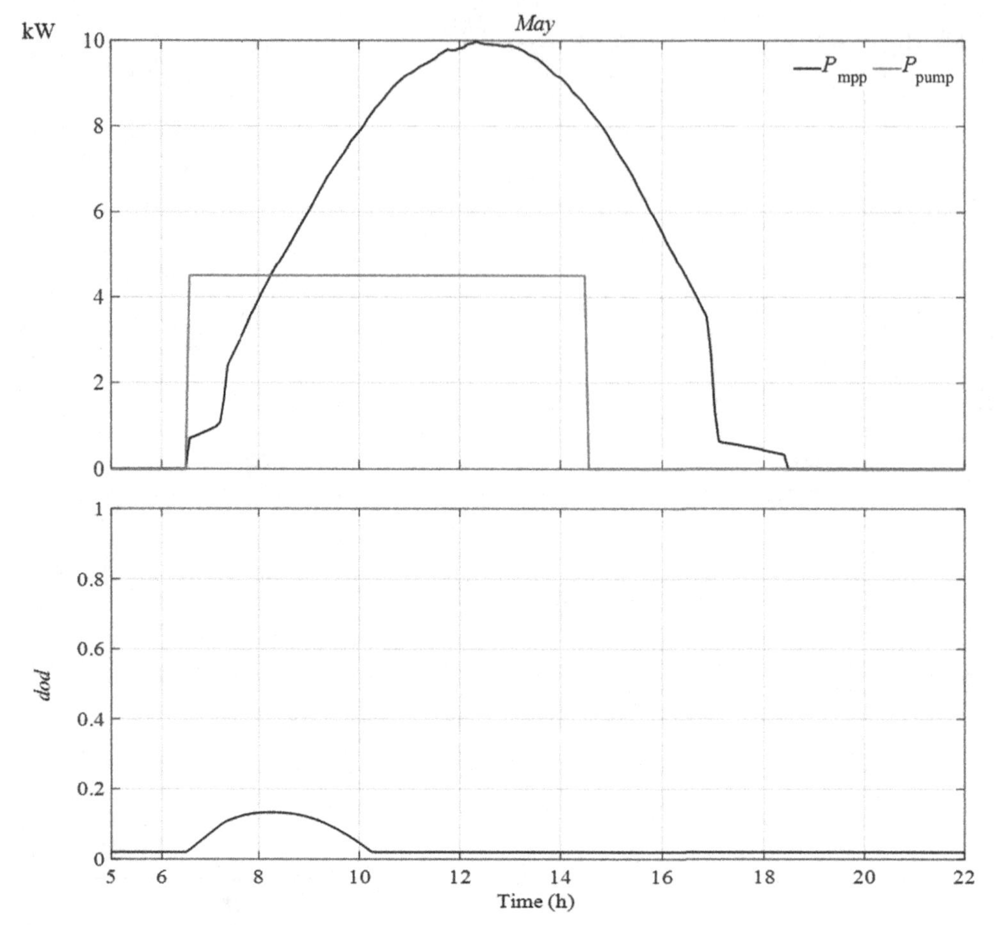

Figure 3.15 Continued.

June and *July*, in which it is equal to 0.962 and 0.934, respectively. This leak of energy can be covered by considering an additional water volume in the reservoir, which will be studied in the next section.

3.4.5 Days of autonomy

In the previous section, the energy installation balance has been demonstrated. In this section, the aim is to evaluate its autonomy. Hence, the case when the cloudy days are consecutive is studied (Table 3.7). The reservoir volume must ensure the installation autonomy. Thus, it must contain the possible leaking water volume in case of consecutive cloudy days and a battery bank totally discharged. Indeed, the water volume in excess or leak is evaluated using Eq. (3.16):

$$V_{\text{leaked/excess}} = V_{\text{pumped}} - \frac{n_{\text{c}}}{f_{\text{i}}} \cdot V \tag{3.16}$$

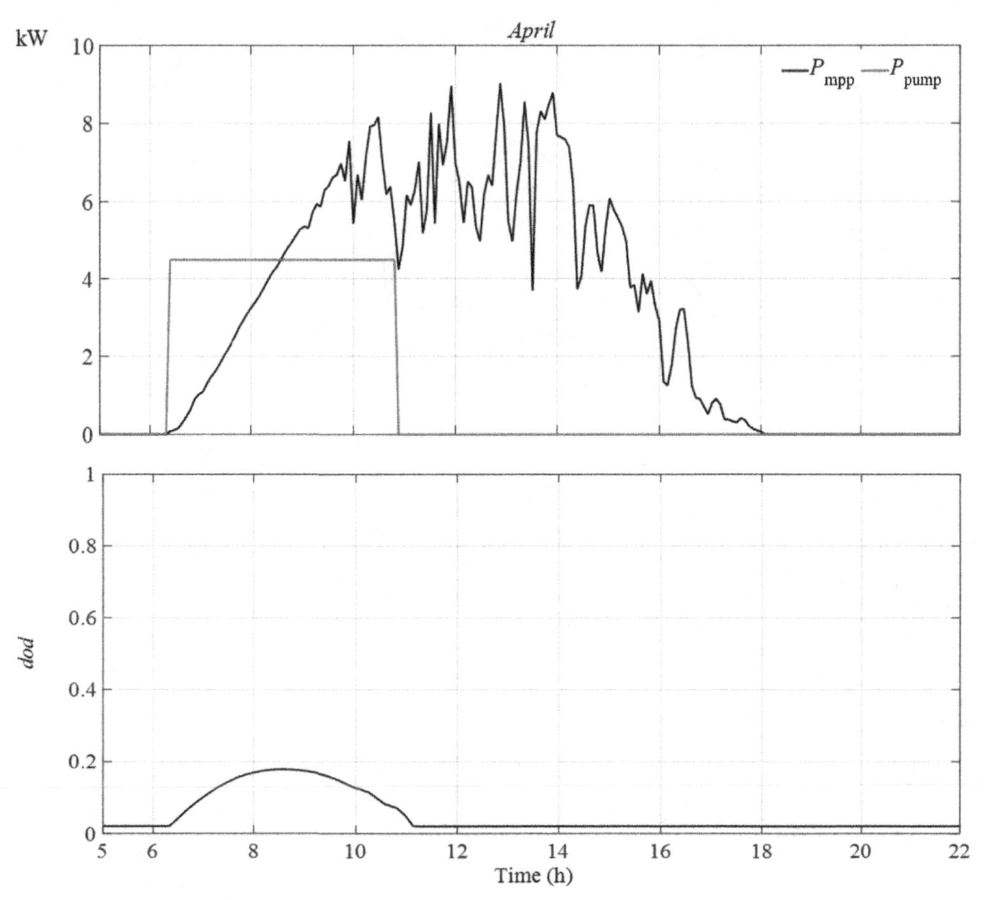

Figure 3.15 Continued.

where:

V_{pumped}: the possible pumped water volume (m^3),

n_c: the number of cloudy days,

f_i: the irrigation frequency,

V: the water volume needed for irrigation for the month M.

Using Table 3.7 and Fig. 3.20, the water volume in leak is maximum in *May* and *July* (1314.6 and 963.13 m^3, respectively). Hence, since these values are close, the volume that corresponds *May* is chosen, to evaluate the reservoir volume. In this case, it is expressed by Eq. (3.17):

$$V_{reservoir} = 1.2 \cdot (V_{leaked/excess} + V) \tag{3.17}$$

Finally, the reservoir volume selected is 1800 m^3.

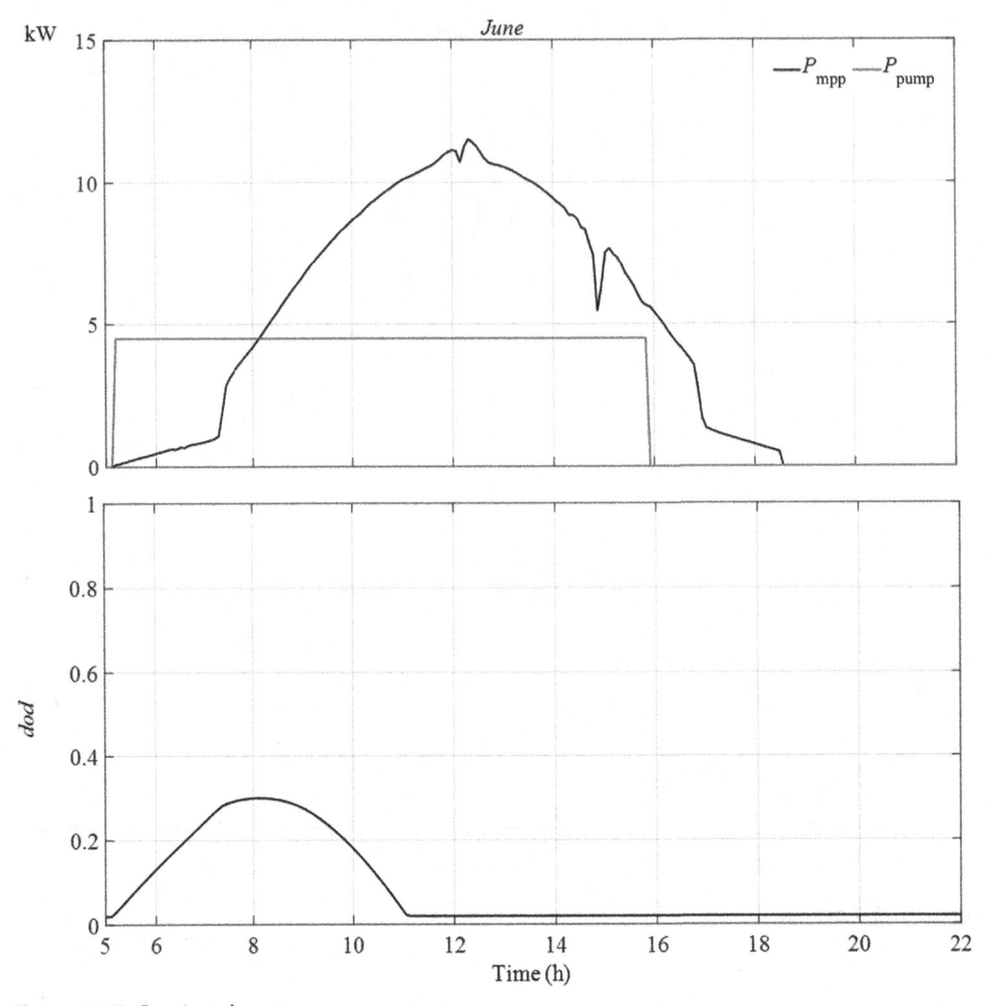

Figure 3.15 Continued.

3.5 CONCLUSIONS

This chapter described the components sizing of an autonomous PV installation for pumping water to irrigate a land planted with tomatoes. A sizing algorithm is proposed to decide on the sizing of the installation elements. The algorithm has been tested for a 10 ha land surface in Medjez El Beb, Tunisia. The results show that the algorithm ensures the system autonomy, the protection of the batteries against deep discharge, and pumping the needed water volume for crops irrigation by ensuring the pump supply during the needed pumping period. The sizing algorithm was tested using measured values of the

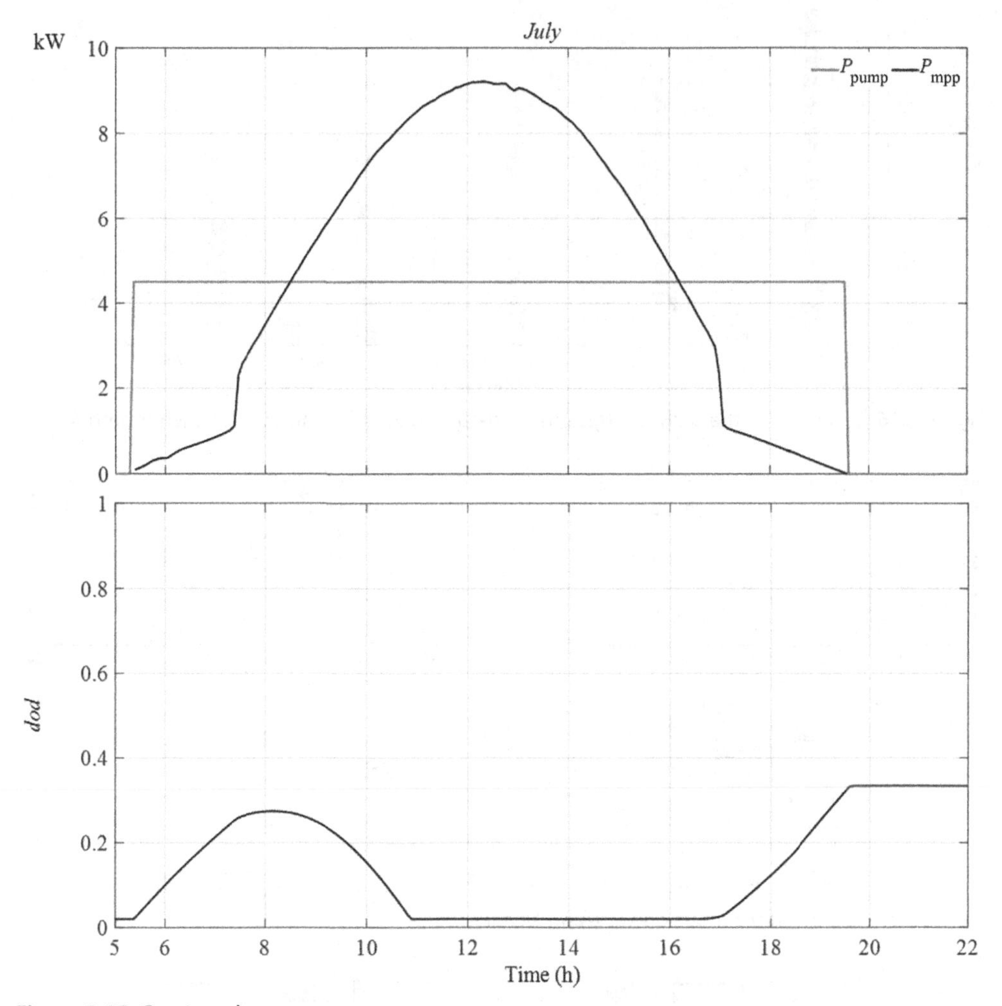

Figure 3.15 Continued.

solar radiation and the ambient temperature of the target area. The obtained results show that the proposed approach gives good results when using measured values. This has been proved by validating the algorithm using HOMER.

As a general conclusion, it has been shown that the proposed algorithm is adequate for the determination of the optimum components sizes for the PV irrigation installation. The results have been presented in Yahyaoui et al. (2013) and Yahyaoui, Ammous, and Tadeo (2015). Using the modeling and sizing results of Chapter 2, Modeling of the Photovoltaic Irrigation Plant Components and this chapter, an online fuzzy management algorithm for the best energy distribution and satisfaction of the requirements of the pump will be described in the next chapter.

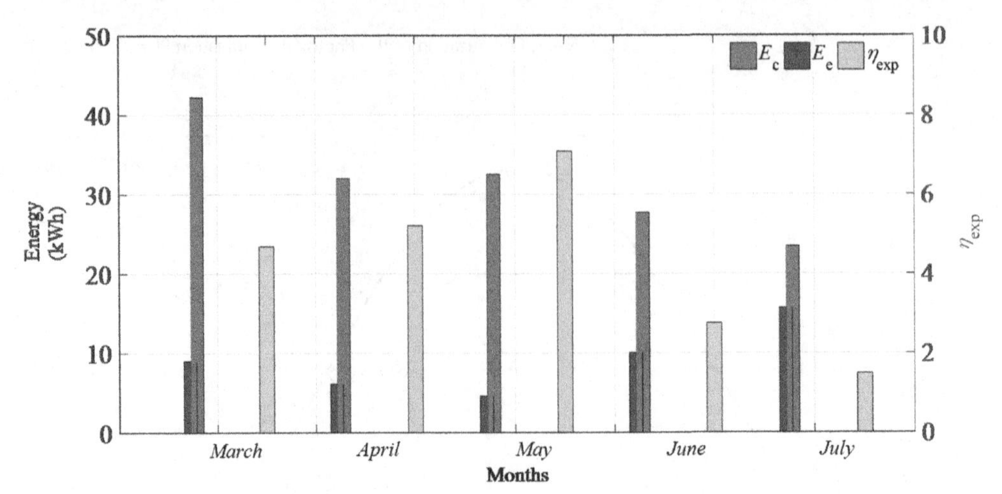

Figure 3.16 Summary of the daily energies using data measured at the 15th of each month *M*.

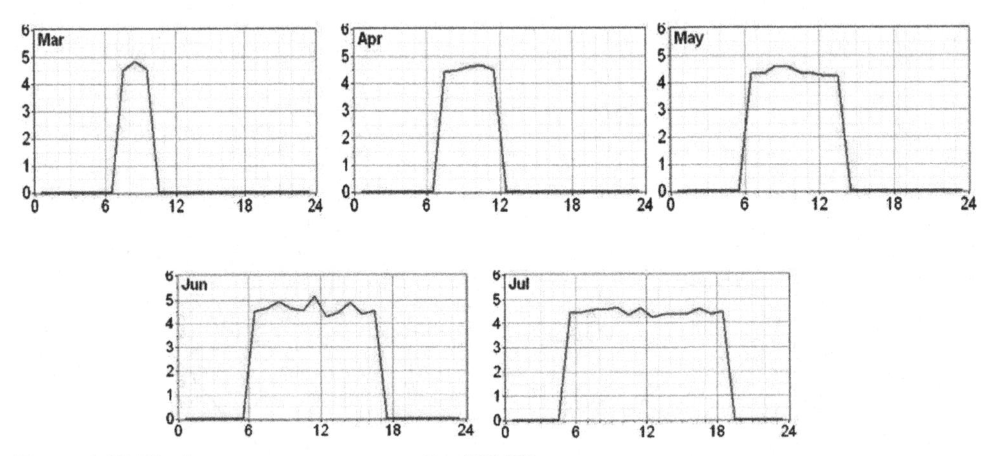

Figure 3.17 The inverter output power using HOMER.

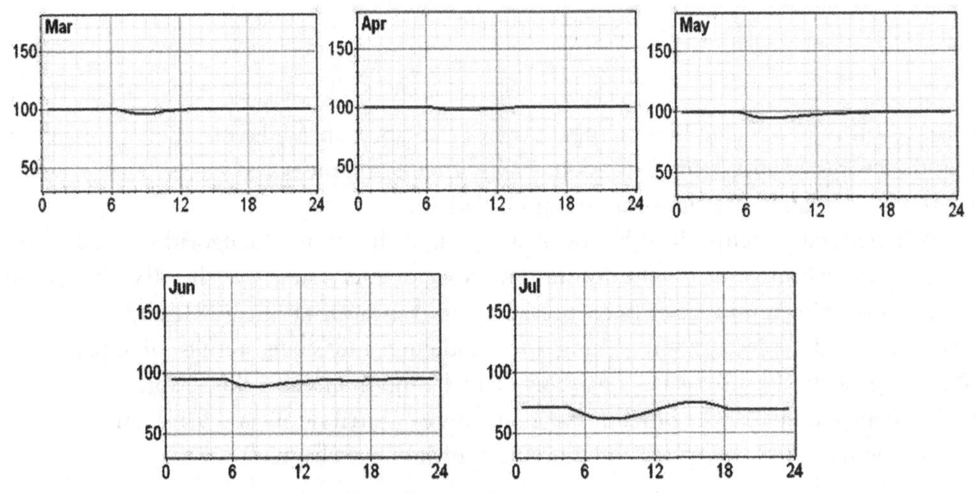

Figure 3.18 The battery bank *SOC* for 1 day in *July*.

PVSYST V5.73 | 24/02/15

Stand Alone System: Simulation parameters

Project : **STAND ALONE PUMPING SYSTEM**

Geographical Site **Medjez** **Country** **Tunisia**

Situation Latitude 35.6°N Longitude 9.6°E

Time defined as Legal Time Time zone UT+1 Altitude 5 m

Albedo 0.20

Meteo data : Medjez, Synthetic Hourly data

Simulation variant : **New simulation variant_final_configuration_Imene**

Simulation date 24/02/15 12h12

Simulation parameters

Collector Plane Orientation Tilt 36° Azimuth 0°

PV Array Characteristics

PV module Si-poly Model **TE55-36P**

Manufacturer Tenesol

Number of PV modules In series 10 modules In parallel 23 strings

Total number of PV modules Nb. modules 230 Unit Nom. Power 55 Wp

Array global power Nominal (STC) **12.65 kWp** At operating cond. 11.33 kWp (50°C)

Array operating characteristics (50°C) U mpp 161 V I mpp 70 A

Total area Module area **102 m²** Cell area 80.6 m²

PV Array loss factors

Thermal Loss factor Uc (const) 20.0 W/m²K Uv (wind) 0.0 W/m²K / m/s

=> Nominal Oper. Coll. Temp. (G=800 W/m², Tamb=20°C, Wind=1 m/s.)NOCT 56 °C

Wiring Ohmic Loss Global array res. 39 mOhm Loss Fraction 1.5 % at STC

Module Quality Loss Loss Fraction 2.3 %

Module Mismatch Losses Loss Fraction 2.0 % at MPP

Incidence effect, ASHRAE parametrization IAM = 1 - bo (1/cos i - 1)bo Parameter 0.05

System Parameter System type **Stand Alone System**

Battery Model **Open 12V / 210 Ah**

Manufacturer Generic

Battery Pack Characteristics Voltage 168 V Nominal Capacity 210 Ah

Nb. of units 14 in series

Temperature Fixed (20°C)

Regulator Model Generic Default with MPPT converter

Technology MPPT converter Temp coeff. -5.0 mV/°C/elem.

Converter Maxi and EURO efficiencies 96.0/94.0 %

Battery Management Thresholds Charging 191.5/183.1 V Discharging 164.6/176.4 V

Back-Up Genset Command 165.5/180.6 V

User's needs : Ext. defined as file load_a_1.txt

Jan.	Feb.	Mar.	Apr.	May	June	July	Aug.	Sep.	Oct.	Nov.	Dec.	Year	
0.00	0.00	419	540	1116	1485	1953	0.00	0.00	0.00	0.00	0.00	5513	kWh

Figure 3.19 Validation of the sizing results using PVsyst.

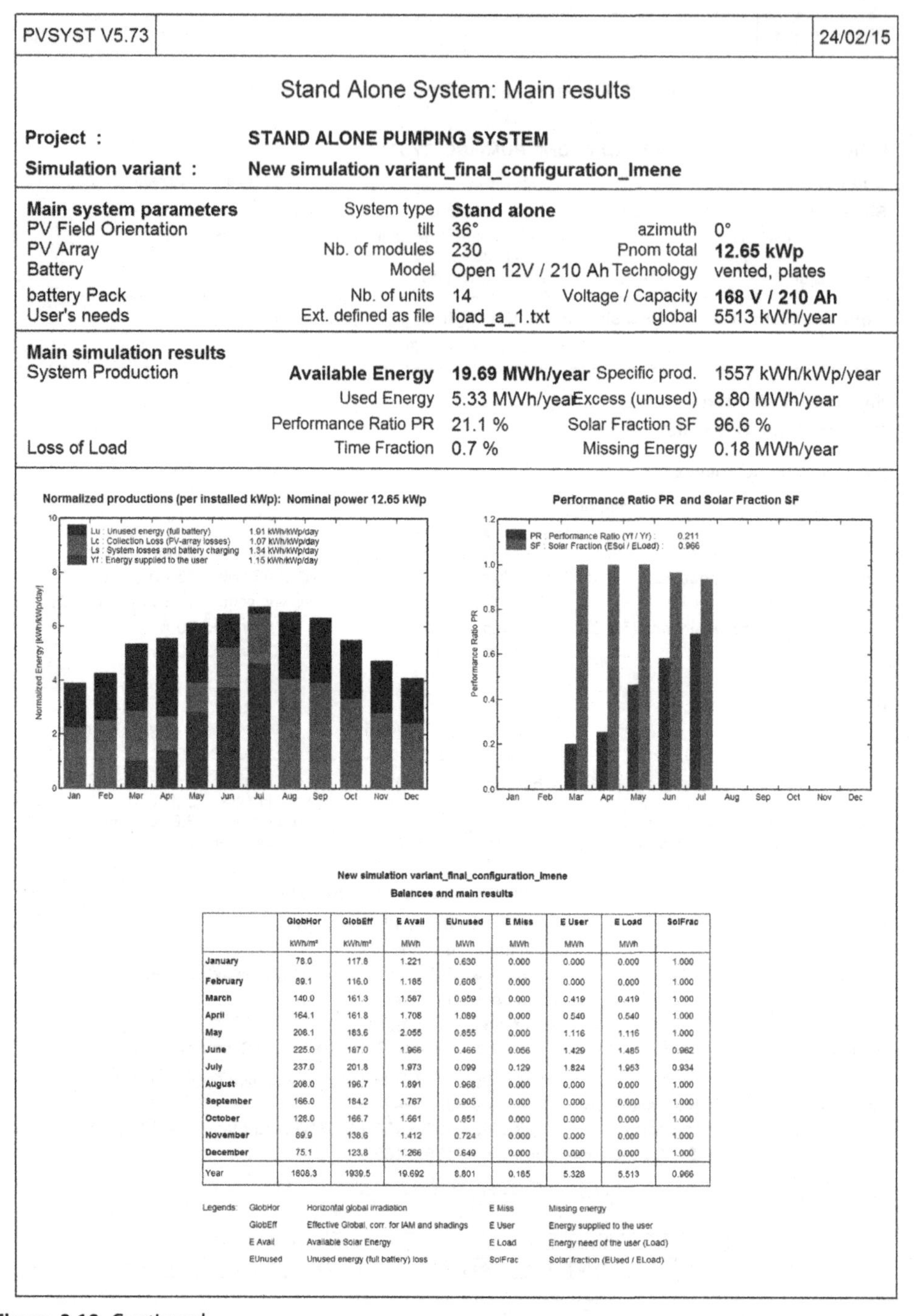

PVSYST V5.73	24/02/15

Stand Alone System: Main results

Project : **STAND ALONE PUMPING SYSTEM**

Simulation variant : **New simulation variant_final_configuration_Imene**

Main system parameters	System type	**Stand alone**		
PV Field Orientation	tilt	36°	azimuth	0°
PV Array	Nb. of modules	230	Pnom total	**12.65 kWp**
Battery	Model	Open 12V / 210 Ah	Technology	vented, plates
battery Pack	Nb. of units	14	Voltage / Capacity	**168 V / 210 Ah**
User's needs	Ext. defined as file	load_a_1.txt	global	5513 kWh/year

Main simulation results				
System Production	**Available Energy**	**19.69 MWh/year**	Specific prod.	1557 kWh/kWp/year
	Used Energy	5.33 MWh/year	Excess (unused)	8.80 MWh/year
	Performance Ratio PR	21.1 %	Solar Fraction SF	96.6 %
Loss of Load	Time Fraction	0.7 %	Missing Energy	0.18 MWh/year

New simulation variant_final_configuration_Imene

Balances and main results

	GlobHor	GlobEff	E Avail	EUnused	E Miss	E User	E Load	SolFrac
	kWh/m²	kWh/m²	MWh	MWh	MWh	MWh	MWh	
January	78.0	117.8	1.221	0.630	0.000	0.000	0.000	1.000
February	89.1	116.0	1.185	0.608	0.000	0.000	0.000	1.000
March	140.0	161.3	1.587	0.959	0.000	0.419	0.419	1.000
April	164.1	161.8	1.708	1.089	0.000	0.540	0.540	1.000
May	208.1	183.6	2.055	0.855	0.000	1.116	1.116	1.000
June	225.0	187.0	1.966	0.466	0.056	1.429	1.485	0.962
July	237.0	201.8	1.973	0.099	0.129	1.824	1.953	0.934
August	208.0	196.7	1.891	0.968	0.000	0.000	0.000	1.000
September	166.0	184.2	1.767	0.905	0.000	0.000	0.000	1.000
October	128.0	166.7	1.661	0.851	0.000	0.000	0.000	1.000
November	89.9	138.6	1.412	0.724	0.000	0.000	0.000	1.000
December	75.1	123.8	1.266	0.649	0.000	0.000	0.000	1.000
Year	1808.3	1939.5	19.692	8.801	0.185	5.328	5.513	0.966

Legends:
- GlobHor: Horizontal global irradiation
- GlobEff: Effective Global, corr. for IAM and shadings
- E Avail: Available Solar Energy
- EUnused: Unused energy (full battery) loss
- E Miss: Missing energy
- E User: Energy supplied to the user
- E Load: Energy need of the user (Load)
- SolFrac: Solar fraction (EUsed / ELoad)

Figure 3.19 Continued.

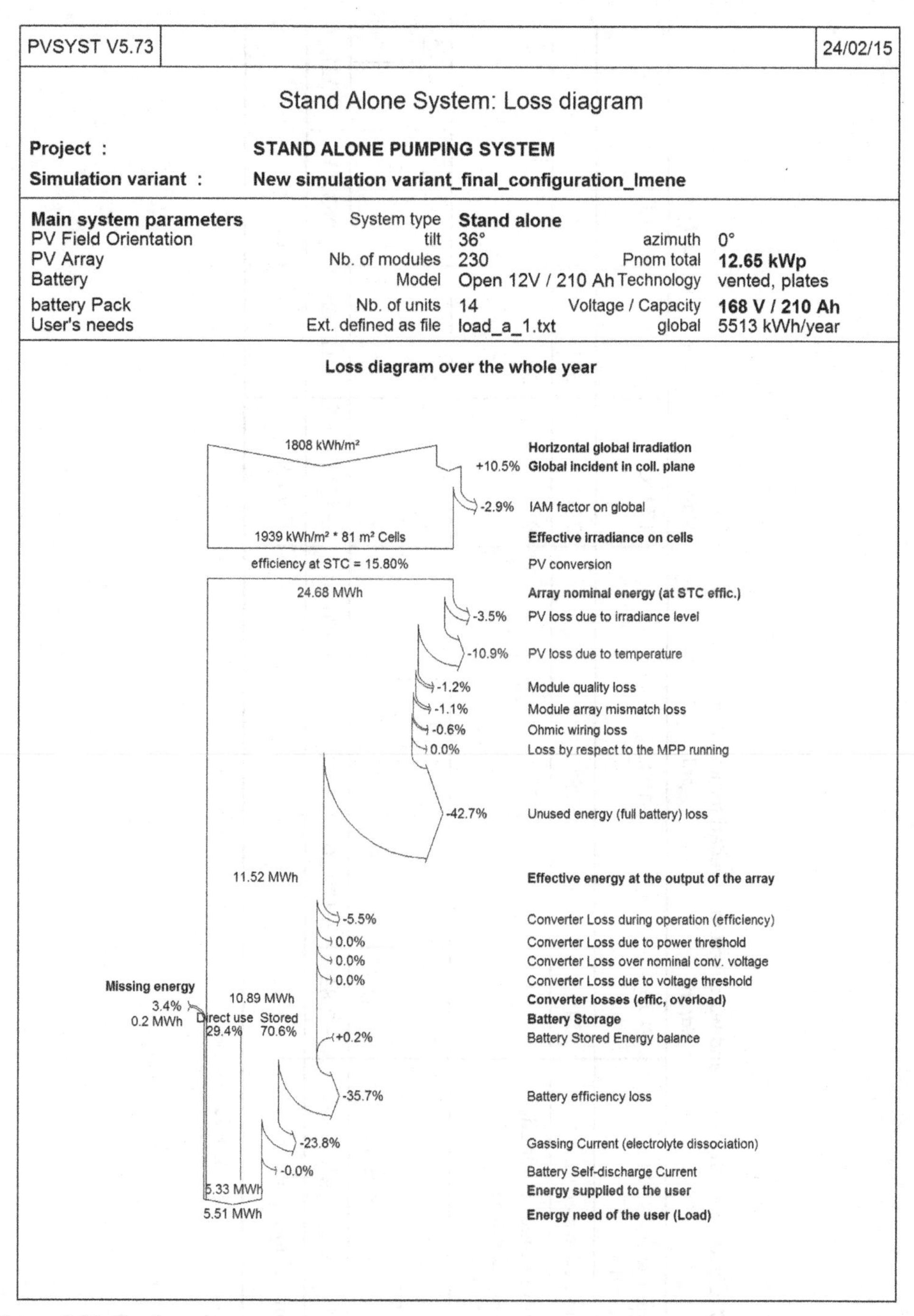

Figure 3.19 Continued.

Table 3.7 Cloudy days frequency and water volume needed for irrigation

<table>
<tr><th>Sizing</th><th colspan="3">March</th><th colspan="3">April</th><th>May</th><th>June</th><th colspan="3">July</th></tr>
<tr><td>Water volume (m^3/10 ha) (Eq. (3.5))</td><td colspan="3">60.70</td><td colspan="3">100.37</td><td>179.82</td><td>241.10</td><td colspan="3">321.03</td></tr>
<tr><td>Daily pumped water (m^3)</td><td colspan="3">274</td><td colspan="3">281.6</td><td>291</td><td>321</td><td colspan="3">321</td></tr>
<tr><td>Maximum number of cloudy days per month n_c (Eq. (3.14))</td><td colspan="3">9</td><td colspan="3">7</td><td>9</td><td>3</td><td colspan="3">4</td></tr>
<tr><td>Irrigation frequency f_i (Chapter: Renewable Energies and Irrigation)</td><td>3</td><td>3</td><td>2</td><td>2</td><td>2</td><td>1</td><td>1</td><td>1</td><td>1</td><td>2</td><td>2</td></tr>
<tr><td>Excess /leak water (m^3)</td><td colspan="2">−7.82</td><td>−129.20</td><td colspan="2">−119.52</td><td>−420.60</td><td>−1314.60</td><td>−417.68</td><td>−963.13</td><td colspan="2">−321.06</td></tr>
<tr><td>Reservoir volume (m^3)</td><td colspan="2"></td><td></td><td colspan="3"></td><td>1793.3</td><td></td><td colspan="3">1541.0</td></tr>
</table>

Bold values denotes Excess/ leak water volume.

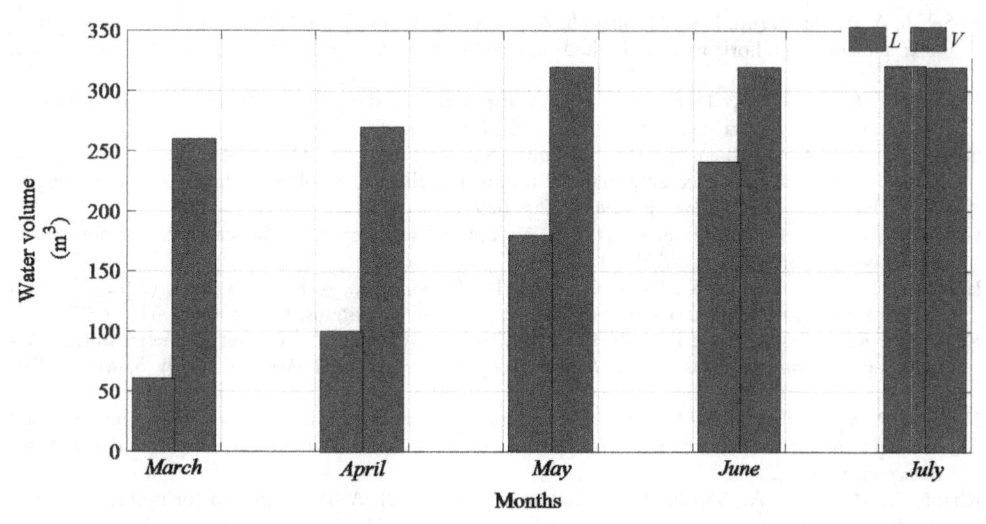

Figure 3.20 Water volume needed for the crops irrigation and pumped using the energy management strategy.

REFERENCES

Abouzahr, I., & Ramakumar, R. (1991). Loss of power supply probability of stand-alone photovoltaic systems: A closed form solution approach. *IEEE Transactions on Energy Conversion, 6*(1), 1–11.

Acakpovi, A.; Xavier, F.F.; & Awuah-Baffour, R. (2012). Analytical method of sizing photovoltaic water pumping system. *Proceedings of the 4th IEEE international conference on adaptive science & technology* (pp. 65–69).

Arunkumar, T., Jayaprakash, R., Denkenberger, D., Ahsan, A., Okundamiya, M. S., Tanaka, H., et al. (2012). An experimental study on a hemispherical solar still. *Desalination, 286*, 342–348.

Barra, L., Catalanotti, S., Fontana, F., & Lavorante, F. (1984). An analytical method to determine the optimal size of a photovoltaic plant. *Solar Energy, 33*(6), 509–514.

Beltrán, J. M. (1999). Irrigation with saline water: Benefits and environmental impact. *Agricultural Water Management, 40*(2), 183–194.

Ben Ammar, M. (2011). Contribution à l'optimisation de la gestion des systèmes multi-sources d'énergies renouvelables (Thesis). Tunisia: National School of Engineering of Sfax.

Ben Salah, C., & Ouali, M. (2012). Energy management of a hybrid photovoltaic system. *International Journal of Energy Research, 36*(1), 130–138.

Bernal-Agustín, J. L., & Dufo-López, R. (2009). Simulation and optimization of stand-alone hybrid renewable energy systems. *Renewable and Sustainable Energy Reviews, 13*(8), 2111–2118.

Chaabene, M. (2009). "Gestion énergétique des systèmes photovoltaïques" (Master course). Tunisia: National School of Engineering of Sfax.

Clean Energy Decision Support Centre. (2001–2004). *"Photovoltaic project analysis, chapter", catalogue no: M39-99/2003E.* RETScreen International, ISBN 0-662-35672-1.

Collares-Pereira, M., & Rabl, A. (1979). The average distribution of solar radiation-correlations between diffuse and hemispherical and between daily and hourly insolation values. *Solar Energy, 22*(2), 155–164.

Eftichios, K., Kolokotsa, D., Potirakis, A., & Kalaitzakis, K. (2006). Methodology for optimal sizing of stand-alone photovoltaic/wind-generator systems using genetic algorithms. *Solar Energy, 80*(9), 1072–1088.

El-Sebaii, A. A., Al-Hazmi, F. S., Al-Ghamdi, A. A., & Yaghmour, S. J. (2010). Global, direct and diffuse solar radiation on horizontal and tilted surfaces in Jeddah, Saudi Arabia. *Applied Energy, 87*(2), 568–576.

Erbs, D. G., Klein, S. A., & Duffie, J. A. (1982). Estimation of the diffuse radiation fraction for hourly, daily and monthly-average global radiation. *Solar Energy, 28*(4), 293–302.

Gong, L., Xu, C.-y, Chen, D. D., Halldin, S., & Chen, Y. D. (2006). Sensitivity of the Penman–Monteith reference evapotranspiration to key climatic variables in the Changjiang (Yangtze River) basin. *Journal of Hydrology, 329*(3), 620–629.

Groumpos, P. P., & Papageorgiou, G. (1987). An optimal sizing method for stand-alone photovoltaic power systems. *Solar Energy, 38*(5), 341–351.

Jakhrani, A. Q., Othman, A.-K., Henry Rigit, A. R., Samo, S. R., & Kamboh, S. A. (2012). A novel analytical model for optimal sizing of standalone photovoltaic systems. *Energy, 46*(1), 675–682.

Kaldellis, J. K., Zafirakis, D., & Kondili, E. (2010). Optimum sizing of photovoltaic-energy storage systems for autonomous small islands. *International Journal of Electrical Power & Energy Systems, 32*(1), 24–36.

Kashyap, P. S., & Panda, R. K. (2001). Evaluation of evapotranspiration estimation methods and development of crop-coefficients for potato crop in a sub-humid region. *Agricultural Water Management, 50*(1), 9–25.

Khatib, T., Mohamed, A., Sopian, K., & Mahmoud, M. (2012). A new approach for optimal sizing of standalone photovoltaic systems. *International Journal of Photo Energy, 2012.*

Khatib, T., Mohamed, Z. A., & Sopian, K. (2013). A review of photovoltaic systems size optimization techniques. *Renewable and Sustainable Energy Reviews, 22*, 454–465.

Klein, S. A., & Beckman, W. A. (1987). Loss-of-load probabilities for stand-alone photovoltaic systems. *Solar Energy, 39*(6), 499–512.

Letey, J., Hoffman, G. J., Hopmans, J. W., Grattan, S. R., Suarez, D., Corwin, D. L., et al. (2011). Evaluation of soil salinity leaching requirement guidelines. *Agricultural Water Management, 98*(4), 502–506.

Maghraby, H. A. M., Shwehdi, M. H., & Al-Bassam, G. K. (2002). Probabilistic assessment of photovoltaic (PV) generation systems. *IEEE Transactions on Power Systems, 17*(1), 205–208.

Mellit, A., Benghanem, M., Hadj Arab, A., & Guessoum, A. (2003). Modelling of sizing the photovoltaic system parameters using artificial neural network. *Proceedings of the IEEE conference on control applications* (pp. 353–357).

Mellit, A., Benghanem, M., & Kalogirou, S. A. (2007). Modeling and simulation of a stand-alone photovoltaic system using an adaptive artificial neural network: Proposition for a new sizing procedure. *Renewable Energy, 32*(2), 285–313.

Moghadam, H., Tabrizi, F. F., & Sharak, A. Z. (2011). Optimization of solar flat collector inclination. *Desalination, 265*(1), 107–111.

Obid, K. R., Khaleel, B., & Nife, K. (2013). The comparison between different methods for estimating consumptive use of water in Iraq. *Journal of Babylon University/Engineering Sciences, 21*, 27–36.

Orgill, J. F., & Hollands, K. G. T. (1977). Correlation equation for hourly diffuse radiation on a horizontal surface. *Solar Energy, 19*(4), 357–359.

Posadillo, R., & López Luque, R. (2009). Hourly distributions of the diffuse fraction of global solar irradiation in Córdoba (Spain). *Energy Conversion and Management, 50*(2), 223–231.

Rana, G., Katerji, N., Lazzara, P., & Ferrara, R. M. (2012). Operational determination of daily actual evapotranspiration of irrigated tomato crops under Mediterranean conditions by one-step and two-step models: Multiannual and local evaluations. *Agricultural Water Management, 115*, 285–296.

Şen, Z. (2008). *Solar energy fundamentals and modeling techniques: Atmosphere, environment, climate change and renewable energy* (Vol. 276). Springer, ISBN 978-1-84800-133-6.

Shrestha, G. B., & Goel, L. (1998). A study on optimal sizing of stand-alone photovoltaic stations. *IEEE Transactions on Energy Conversion, 13*(4), 373–378.

Sidrach-de-Cardona, M., & Mora López, L. (1998). A simple model for sizing stand-alone photovoltaic systems. *Solar Energy Materials and Solar Cells, 55*(3), 199–214.

Van Hoorn, J. W. (1981). Salt movement, leaching efficiency, and leaching requirement. *Agricultural Water Management*, *4*(4), 409–428.

Yahyaoui, I., Ammous, M., & Tadeo, F. (2015). Algorithm for optimum sizing of a photovoltaic water pumping system. *International Journal of Computer Applications (IJCA)*, *11*(6), 21–28.

Yahyaoui, I., Chaabene, M., & Tadeo, F. (2013). An algorithm for sizing photovoltaic pumping systems for tomato irrigation. *Proceedings of the IEEE conference on renewable energy research and applications (ICRERA)* (pp. 1089–1095).

Yahyaoui, I., Tadeo, F., & Segatto, M. E. V. (2016). Energy and water management for drip-irrigation of tomatoes in a semi-arid district. *Agricultural Water Management*.

Yang, H., Zhou, W., Lu, L., & Fang, Z. (2008). Optimal sizing method for stand-alone hybrid solar–wind system with LPSP technology by using genetic algorithm. *Solar Energy*, *82*(4), 354–367.

Zhang, C. P., Sharkh, S. M., Li, X., Walsh, F. C., Zhang, C. N., & Jiang, J. C. (2011). The performance of a soluble lead-acid flow battery and its comparison to a static lead-acid battery. *Energy Conversion and Management*, *52*(12), 3391–3398.

CHAPTER 4

Optimum Energy Management of the Photovoltaic Irrigation Installation

4.1 INTRODUCTION

As it has been seen in Chapter 2, Modeling of the Photovoltaic Irrigation Plant Components, PVPs generate energy intermittently, due to the frequent changes in the solar radiation (Abdeen, 2008; Singh, 2013; Jung & Ahmed, 2012) (see Section 2.3). Moreover, the energy generation depends on the climatic parameters (namely, the solar radiation), the site, and the panel characteristics (Celik & Acikgoz, 2007; Etienne, Alberto, & Mikhail, 2011; Lo Brano, Orioli, Giuseppina, & Di Gangi, 2010). In general, PV energy is abundant during the warm season, characterized by an important sunlight, and it is low during the cold season, with rapid changes in the solar radiation. Hence, due to the variability of the PV energy generated, the use of an energy management strategy is important, especially for autonomous installations, for which a precise balance between the PV power generated and the power demanded by the load is required (Gergaud, 2002; Sallem, Chaabene, & Kamoun, 2009a, 2009b; Semaoui, Hadj Arab, Bacha, & Azoui, 2013).

In this context, this chapter focus on the energy management of the off-grid PV installation (Fig. 4.1) presented in Chapters 2 and 3, Modeling of the Photovoltaic Irrigation Plant Components; Sizing Optimization of the Photovoltaic Irrigation Plant Components, destined to supply the water pump for irrigation. In fact, as the system is off-grid, the installation must always provide the energy needed by the load. Thus, in case of insufficient generated energy, a battery bank is used to provide the missing energy (Fig. 4.1). The main objective of the management approach, developed in this chapter, is to decide the connection of the system components that ensures supplying the load and pumping the water volume needed, taking into account the safe operation of the battery bank. Hence, a careful planning of the load connections to the PV generator and/or to the battery bank is required. Based on optimization tools, the proposed strategy guarantees a maximum use of the energy generated by the PVPs.

Thus, this chapter establishes a management strategy for a PV installation destined to water pumping for irrigation of a land planted with tomatoes (Yahyaoui, Chaabene, & Tadeo, 2014a). Section 4.2 reviews some methods used in the literature for the energy management of renewable irrigation installations. Then, Section 4.3

Specifications of Photovoltaic Pumping Systems in Agriculture
DOI: http://dx.doi.org/10.1016/B978-0-12-812039-2.00004-1

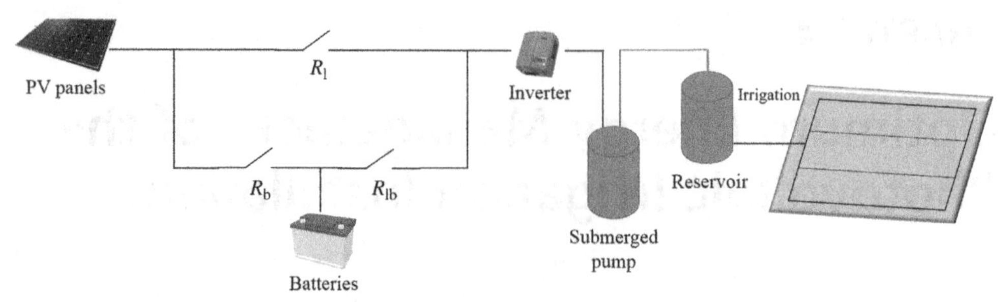

Figure 4.1 Scheme of the off-grid PV irrigation system.

presents the problem formulation. The energy management approach, based on fuzzy logic, is presented in Section 4.4. The Energy Management Algorithm (EMA) is then tested on a case study in Section 4.5 and validated experimentally in Section 4.6. Finally, Section 4.7 presents the chapter conclusion.

4.2 REVIEW OF RENEWABLE ENERGY MANAGEMENT IN IRRIGATION

As it has been mentioned, the energy management is important in off-grid applications, especially for installations for which autonomy is required, independently of the climatic conditions. Hence, the choice of the adequate management method is relevant in optimizing the energy exploitation. Some of these methods are discussed below.

4.2.1 Review of energy management methods

In the literature, several management methods have been studied for PV water pumping installations. In fact, some works have focused in pumping water over the sun (Arrouf & Ghabrour, 2007; Belgacem, 2012; Betka & Attali, 2010; Ghoneim, 2006). These installations directly use all the PV energy generated for pumping water, so they are simple, but water is only pumped when there is enough solar radiation. To solve this, PV installations equipped with batteries are used: The regulation of the battery bank voltage has been used to disconnect or connect loads to both PVPs and batteries, and thus charging or discharging the energy storage devices (Messikh, Chikhi, Chikhi, & Chergui, 2008). This method allows the battery bank voltage to be regulated. However, the battery bank is always operating, which reduces its lifetime. In addition, intelligent tools have been used for the energy management. For instance, ANNs have been developed for the optimum operation of water pumping installations (Kalogirou, 2001). Moreover, the efficiency of Genetic Algorithm (GA) and nonlinear programming has been proved for energy management of PV irrigation installations (Al-Alawi, Al-Alawi, & Islam, 2007; Moradi-Jalal, Marino, & Afshar, 2003). Despite the efficiency of these methods, the creation and updating of the required database remains the main drawback of the ANN (Hemanth, Vijila, & Anitha, 2010; Liu, Fang, Qin, Ye, & Xie, 2011), while the main disadvantage of GA is its long counting time and the lack of guarantee that a global optimal solution can always be found (Lee, Lin, Liao, & Tsao, 2011; Moradi & Abedini, 2012). For this purpose, fuzzy logic has been extensively used (Ben Salah, Chaabene, & Ben Ammar, 2008; Berrazouane & Mohammedi, 2014; Bouchon-Meunier, Foulloy, & Ramdani, 1994; Cheikh, Aïssa, Malek, & Becherif, 2010; Hahn, 2011; Kahraman, Kaya, & Cebi, 2009; Saad & Arrofiq, 2012; Welch & Venayagamoorthy, 2010; Yahyaoui, Chaabene, & Tadeo, 2014b; Yahyaoui, Sallem, Kamoun, & Tadeo, 2014). Hence, this approach is selected here and reviewed in detail.

4.2.2 Fuzzy logic for energy management

Fuzzy logic has already proven its efficiency in ensuring an adequate energy management of PV plants (Ben Salah et al., 2008). In fact, in autonomous PV installations, this tool has been used for optimizing the energy use, guaranteeing the plant autonomy, and protecting the batteries against deep discharge and excessive charge, while supplying noncontrollable loads (Yahyaoui et al., 2014). Moreover, it has been used in renewable installations that supply controllable loads (Ben Salah et al., 2008; Berrazouane & Mohammedi, 2014) and constant critical and time-varying noncritical loads (Welch & Venayagamoorthy, 2010). Researchers have also used it in simpler applications in agriculture. For instance, fuzzy logic has been used to decide crop irrigation time and nutrient injection, knowing the solar radiation and the crop canopy temperature (Hahn, 2011). Moreover, the efficiency of this tool has been proven in controlling the internal climatic variables in greenhouses and batteries charging/discharging (Cheikh et al., 2010), and increasing the pumped water volume in water pumping installations, using Fuzzy Management Algorithm (FMA) (Sallem et al., 2009a, 2009b).

The efficiency of this control method in various applications is given by its ease of use. In fact, it is complicated in many installations to give exact rules for energy management (Ben Salah et al., 2008; Sallem et al., 2009a). Thus, fuzzy logic is considered a good method for solving this problem, since it gives a simple method to decide control actions, using linguistic rules (Kahraman et al., 2009; Sallem et al., 2009b). Moreover, based on the knowledge base, the fuzzy rules are written in a simple manner that describes directly the control decisions. Mamdani-type fuzzy logic is used here within the management algorithm, as it is simple to learn for operators with little technical training (Bouchon-Meunier et al., 1994) and can be implemented using standard components, such as programmable industrial controllers (Saad & Arrofiq, 2012). For these reasons, in the application studied in this book, fuzzy logic is chosen for the energy management approach of the PV irrigation installation (Yahyaoui et al., 2014a, 2014b).

4.3 PROBLEM FORMULATION

The proposed energy management strategy aims to determine, at each sample time, the electrical energy produced from a PV installation composed of a PV generator and a battery bank, which supply a water pump (Fig. 4.1). More precisely, the goal is to develop a management algorithm that maximizes the use of PV power generated, minimizes the battery use, and guarantees the water volume needed for irrigation, by controlling the switching of the relays R_b, R_l, and R_{lb} that link the installation components (Fig. 4.1).

The decision on these switchings is carried by a management algorithm, based on the estimations of the PV power generated $\tilde{P}_{pv}$, the power demanded by the pump $\tilde{P}_{pump}$, the battery bank *dod*, and the water volume in the reservoir L. $\tilde{P}_{pv}$ is estimated using the measured values of the solar radiation G and the ambient temperature T_a. $\tilde{P}_{pump}$ and *dod* are estimated using the measured currents I_{pump} and I_{bat}, respectively (see Section 2.3). The level L is measured directly using a pressure sensor.

The proposed energy management is then performed via two main steps: The first step consists in the acquisition of the currents I_{pv} and I_{bat}, and the acquisition or the prediction of the climate-related site parameters; this allows the power generated by the PVPs $\tilde{P}_{pv}$ and the battery bank' *dod* to be predicted. In the second step, using the fuzzy logic, an Energy Management Algorithm (EMA) deduces the instants and duration when the load is connected to the power sources (Fig. 4.2).

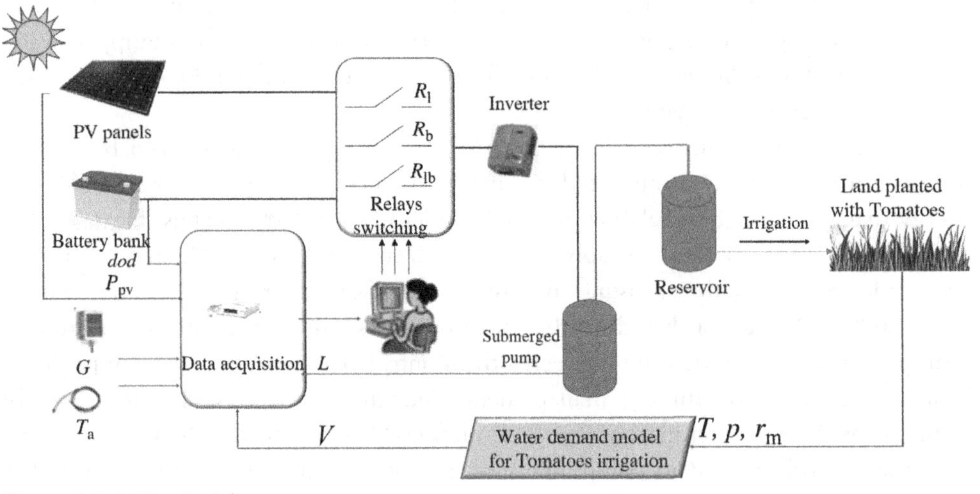

Figure 4.2 EMA principle.

4.4 PROPOSED EMA

As it has been mentioned in Section 4.1, the aim of this chapter is to develop an algorithm for the efficient energy management of the autonomous PV irrigation installation, composed of PVPs coupled to a lead–acid battery bank, to ensure the energy availability, even if the irradiation is low (Fig. 4.1). Based on some previous proposals (Sallem et al., 2009a; Yahyaoui et al., 2014a), a general FMA is proposed here.

4.4.1 Energy management strategy

The aim is to establish a management algorithm that ensures the water volume needed for the crops irrigation, through the control of the relays that link the installation components. Hence, an FMA is proposed here to decide the connection time of the system components, using only a knowledge base of the system (Sallem et al., 2009b) (Section 4.2). The components connection times are decided by means of fuzzy rules, which are based on the estimated PV power and the water volume in the reservoir, while taking into account the constraints related to the battery bank safe operation.

In normal cases, the PVPs are used to supply the pump and charge the battery bank. To minimize the battery use, the water pumping is performed during the daylight. This ensures a *dod* between two fixed values dod_{min} and dod_{max} for a continuous pump operation (that stops when the tank is full or the battery discharged).

The management algorithm decides the switching times of the three relays R_b, R_l, and R_{lb} that connect the PV system components (Fig. 4.1). Hence, it is necessary to establish some criteria that define the algorithm. These criteria are related to (Fig. 4.3):

- the PV power generated P_{pv},
- the battery bank *dod*,
- the water volume in the reservoir *L*.

The management criteria are then defined as follows:

- Maintain a high water level in the reservoir, to guarantee the water volume needed for the crop irrigation.
- When the reservoir contains enough water, store the excess of PV energy in the battery bank.
- Ensure a *dod* less than dod_{max}, to protect the batteries against the deep discharge, and higher than dod_{min}, to protect them from the excessive charge.
- Ensure a margin of 10% of the PV power: The pump can be connected to the panel only if the measured PV power P_{pv} is 10% higher than the required power by the pump P_{pump}, to guarantee a continuous power supply for the pump.

These criteria are established to meet fixed objectives presented in Fig. 4.3:

O1. Provide the required irrigation when needed, by storing water in the reservoir.

O2. Minimize the use of the battery bank.

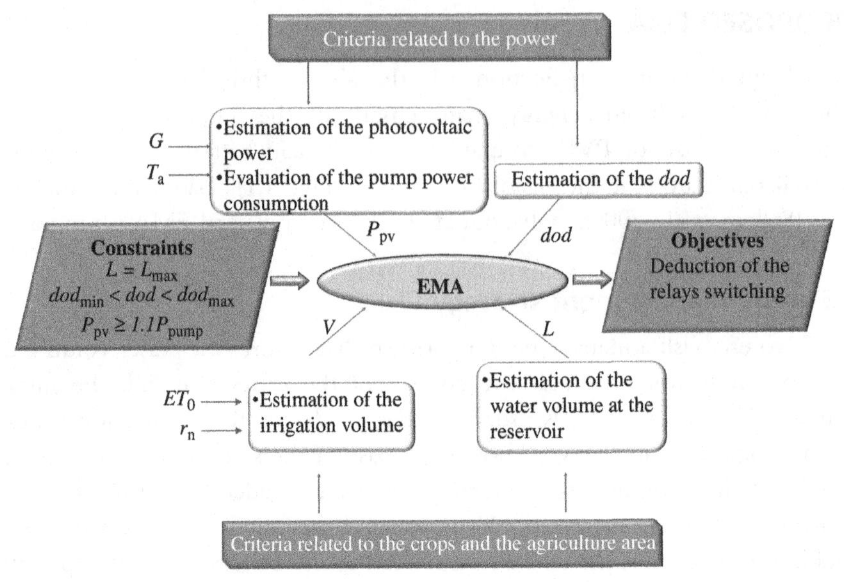

Figure 4.3 Structure of the proposed EMA.

O3. Protect the batteries against the excessive charge and discharge, by disconnecting them, respectively, from PVs and the pump when they are not used.

O4. Ensure a continuous power supply, especially during weather changes:

During the day, the instantaneous power P_{pump} verifies that:

$$P_{pump} = P_{pv} + \overline{P}_{bat} \tag{4.1}$$

and the current absorbed by the load I_{pump} is just:

$$I_{pump} = I_{pv} + \overline{I}_{bat} \tag{4.2}$$

According to the fourth criterion, the panels supply directly the load without any contribution from the battery bank when they provide at least 110% of the demand. This criterion ensures a continuous power supply of the pump. Thus:

$$I_{pv} \geq 1.1\, I_{pump} \tag{4.3}$$

As it has been mentioned, in order to derive an EMA that can be easily implemented, tuned, maintained by persons with low technical training, and adapted to different installations and irrigation facilities, the management algorithm is implemented using fuzzy logic.

The proposed algorithm needs preliminary treatment of some data that will be provided later to the fuzzy algorithm: the expected PV power (using the model given by Campana, Li, & Yan, 2013), the *dod* (Sallem et al., 2009a, 2009b), and the stored water in the reservoir. In fact, during the night, in case of lack of water volume in the reservoir, due to an unexpected water extraction or in case of pumping insufficient water volume during the day, the missing water volume is recovered using the battery bank, if it is charged.

For this, the length of time Δt_{bat} for which the battery is capable to supply the pump without exceeding dod_{max} is evaluated using Eq. (3.5) (Sallem et al., 2009a, 2009b):

$$\Delta t_{bat} = (dod_{max} - dod(t))\frac{C_P}{I_{bat}^{k_P}} \tag{4.4}$$

where:

dod_{max}: the maximum allowed value for the *dod*,

C_P: the Peukert capacity (Ah),

I_{bat}: the battery current (A),

k_P: the Peukert constant.

After estimating these inputs, the fuzzy tool generates the decisions, following the management algorithm that is now explained.

4.4.2 Relays switching modes

To achieve the objectives O1–O4 explained in Section 4.4.1, six operating modes are defined for the switching of the three relays R_b, R_l, and R_{lb}:

1. At night, in normal conditions, the volume in the reservoir is full, so all the relays would be *off* (*mode 1*). This mode would be maintained during the irrigation period, when the tank volume decreases.
2. During the early hours of the morning, *mode 2* is possible, since the battery bank and the PVPs supply the pump. In this case, the relays R_l and R_{lb} would be *on*.
3. *Mode 3* consists in pumping water and charging the battery bank with the energy in excess. In this case, the relays R_l and R_b would be *on* and the relay R_{lb} would be *off*.
4. When the reservoir is full, the PV energy generated is fully used to charge the battery bank when the batteries are discharged. This corresponds to *mode 4*, for which the relay R_b would be *on*.
5. The relay R_l is switched *on* during the fifth mode (*mode 5*), to allow the pump to be supplied when the panels produce a sufficient power to the pump, with 10% margin.
6. During *mode 6*, only the relay R_{lb} would be switched *on*. This mode is possible during the night when the water volume in the reservoir is less than the volume needed to irrigate the crops.

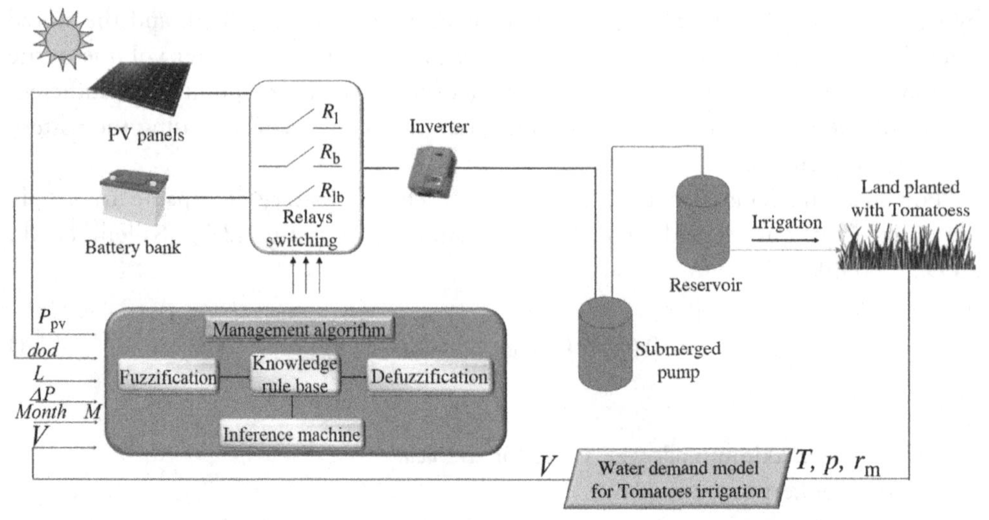

Figure 4.4 Proposed structure for the implementation of the management system.

4.4.3 Fuzzy EMA

The EMA is based on four steps: the extraction of the knowledge base, the fuzzification, the inference diagram, and the defuzzification (Beccali, Cellura, & Ardente, 1998; Berrazouane & Mohammedi, 2014). These four steps are presented now in detail in this proposal, following the structure presented in Fig. 4.4.

4.4.3.1 *Knowledge base*

The knowledge base is generated on the basis of the specifications:

PV power $\tilde{P}_{pv}$

The PV power $\tilde{P}_{pv}$ generated is periodically estimated and then partitioned in three fuzzy sets that cover the interval $X = [0, P_{pv\ max}]$ at *low, medium*, and *high* power generation levels, respectively:

$$\forall x \in \mathcal{X} \ \mu_L(x) + \mu_M(x) + \mu_H(x) = 1 \tag{4.5}$$

where $\mu_L(x)$, $\mu_M(x)$, and $\mu_H(x)$ are, respectively, the *low, medium*, and *high* membership functions at the measured power level x.

Battery *dod*

It is composed of three fuzzy sets that cover the interval $D = [0, dod_{max}]$ at *low, medium*, and *high* production levels, respectively, and verify:

$$\forall d \in \mathcal{D}, \ \mu_{dL}(d) + \mu_{dM}(d) + \mu_{dH}(d) = 1 \tag{4.6}$$

where $\mu_{dL}(d)$, $\mu_{dM}(d)$, and $\mu_{dH}(d)$ are, respectively, the *low*, *medium*, and *high* membership functions of the estimated *dod* d.

Stored water volume ν

The third partition is composed of three fuzzy sets in the interval $V = [0, V_{\max}]$ which verify:

$$\forall \nu \in \mathcal{V}, \ \mu_{\nu L}(\nu) + \mu_{\nu M}(\nu) + \mu_{\nu H}(\nu) = 1 \tag{4.7}$$

where $\mu_{\nu L}(\nu)$, $\mu_{\nu M}(\nu)$, and $\mu_{\nu H}(\nu)$ are, respectively, the membership functions of ν.

As the definition of *low*, *medium*, and *high* depends on the use of the auxiliary sets, the following fuzzy variables are defined:

Months *M* This partition is composed of as many fuzzy sets as months, given by the set $\mathcal{M} = \{\mathbf{m_1}, \mathbf{m_2}, \ldots, \mathbf{m_t}\}$ and verify:

$$\forall m \in \mathcal{M}, \mu_{m_1}(m) + \mu_{m_2}(m) + \cdots + \mu_{m_t}(m) = 1 \tag{4.8}$$

where $\mu_{m_i}(m)$ are the membership functions corresponding to the month m.

Water level *L* This partition is composed of as many fuzzy sets as months, denoted by the set $\mathcal{L} = (l_1, l_2, \ldots, l_t)$. The interval of the possible water level $L = [0, L_{\max}]$ is covered by these fuzzy sets and verify:

$$\forall l \in \mathcal{L} \ \mu_{l_1}(l) + \mu_{l_2}(l) + \cdots + \mu_{l_t}(l) = 1 \tag{4.9}$$

where $\mu_{l_i}(l)$ is the membership function corresponding to l_i evaluated at l.

Power difference ΔP

This partition is composed of two fuzzy sets $\mathcal{F} = (f_1, f_2)$ and verify:

$$\forall f \in \mathcal{F}, \ \mu_{f_1}(f) + \mu_{f_2}(f) = 1 \tag{4.10}$$

where $\mu_{f_e}(f)$ is the membership function corresponding to f_e evaluated at f.

Relays R_l, R_b, R_{lb}

To decide the switching of the relays R_l, R_b, R_{lb}, depending on the fuzzy variables x, d, and ν, two fuzzy sets are planned $O = \{\text{on}, \text{off}\}$. They cover the domain $O = [0, 1]$ and verify $\forall o \in \mathcal{O}$:

$$\begin{cases} \mu_{\text{off } r_l}(o) + \mu_{\text{on } r_l}(o) = 1 \\ \mu_{\text{off } r_b}(o) + \mu_{\text{on } r_b}(o) = 1 \\ \mu_{\text{off } r_{lb}}(o) + \mu_{\text{on } r_{lb}}(o) = 1 \end{cases} \tag{4.11}$$

where the switching controls given to relays are provided by the membership functions corresponding to r_l, r_b, r_{lb}, respectively, evaluated at o.

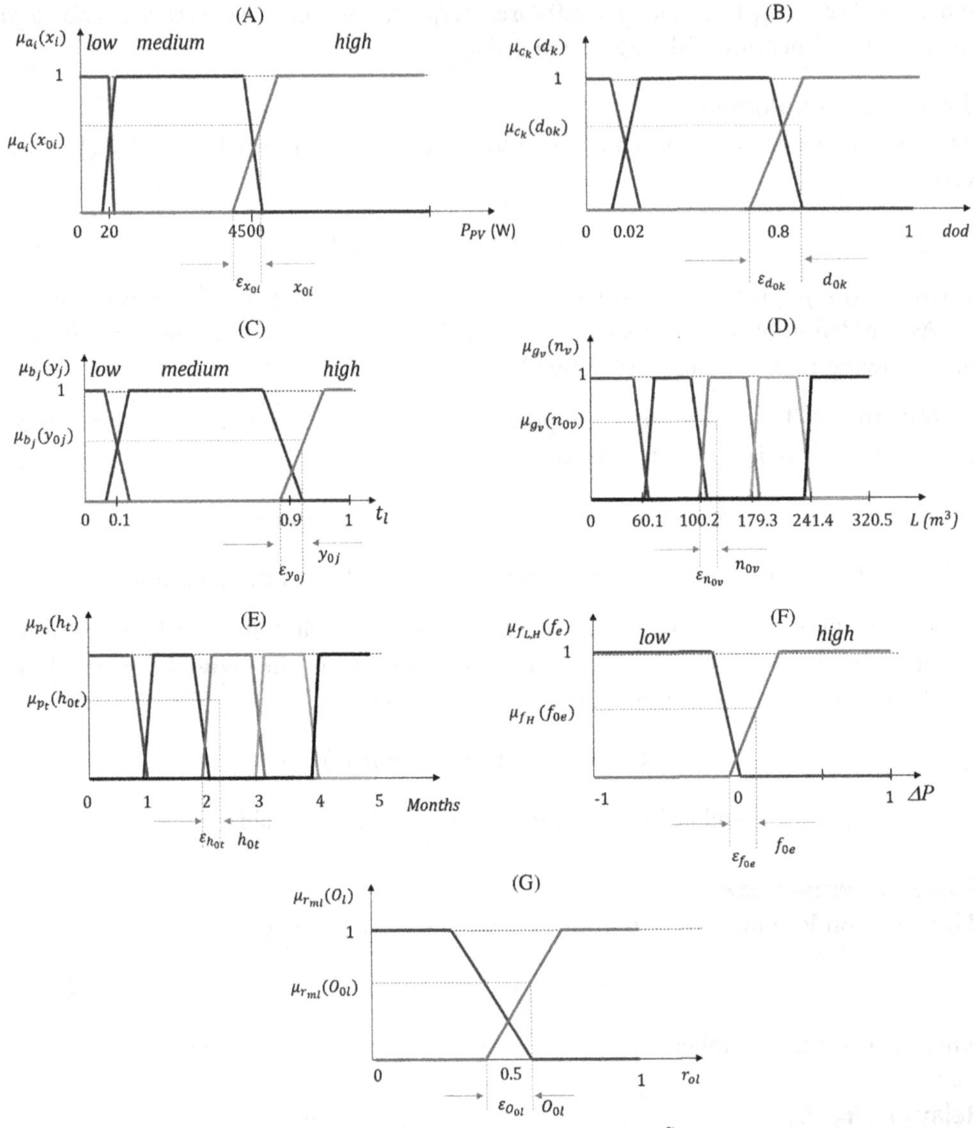

Figure 4.5 Membership functions corresponding to: (A) PV power $\tilde{P}_{pv}$, (B) battery *dod*, (C) fuzzified water volume *v*, (D) water volume *L*, (E) months *M*, (F) power difference ΔP, and (G) control signals of each relay O_z

Based on this structure, the fuzzy rules of the relays switching times are classified according to three sets of *dod* (Fig. 4.5):

- $dod \in [0\ d_{dL_{\max}}]$: the pump is supplied by the panels and/or the battery bank,
- $dod \in [d_{dL_{\min}}\ d_{dM_{\max}}]$: supplying the pump is preferred to charging the battery bank,
- $dod \in [d_{dM_{\min}}\ d_{dH_{\max}}]$: charging the battery bank is preferred to supplying the pump since the panels produce insufficient power to the pump and the battery bank is discharged.

4.4.3.2 *Fuzzification*

PV Power $\tilde{P}_{pv}$

The membership functions of $\mu_L(x_{0i})$, $\mu_M(x_{0i})$, $\mu_H(x_{0i})$ corresponding to $\tilde{P}_{pv}$ are expressed as follows (Fig. 4.5):

$$\mu_L(x_{0i}) = \begin{cases} 1 & \text{if} \quad 0 < x < x_{L_{\min}} \\ \dfrac{x_{0i} - x}{\varepsilon_{x_{0i}}} & \text{if} \quad x_{L_{\min}} < x < x_{L_{\max}} \\ 0 & \text{otherwise} \end{cases} \tag{4.12}$$

$$\mu_M(x_{0i}) = \begin{cases} \dfrac{x - x_{0i}}{\varepsilon_{x_{0i}}} & \text{if} \quad x_{M_{\min 1}} < x < x_{M_{\min 2}} \\ 1 & \text{if} \quad x_{M_{\min 2}} < x < x_{M_{\max 1}} \\ \dfrac{x_{0i} - x}{\varepsilon_{x_{0i}}} & \text{if} \quad x_{M_{\max 1}} < x < x_{M_{\max 2}} \\ 0 & \text{otherwise} \end{cases} \tag{4.13}$$

$$\mu_H(x_{0i}) = \begin{cases} 1 & \text{if} \quad x > x_{H_{\max}} \\ \dfrac{x - x_{0i}}{\varepsilon_{x_{0i}}} & \text{if} \quad x_{H_{\min}} < x < x_{H_{\max}} \\ 0 & \text{otherwise} \end{cases} \tag{4.14}$$

Battery *dod*

The membership functions of $\mu_{dL}(d_{0k})$, $\mu_{dM}(d_{0k})$, $\mu_{dH}(d_{0k})$ corresponding to *dod* are expressed as follows (Fig. 4.5):

$$\mu_{dL}(d_{0k}) = \begin{cases} 1 & \text{if} \quad 0 < d < d_{dL_{\min}} \\ \dfrac{d_{0k} - d}{\varepsilon_{d_{0k}}} & \text{if} \quad d_{dL_{\min}} < d < d_{dL_{\max}} \\ 0 & \text{otherwise} \end{cases} \tag{4.15}$$

$$\mu_{dM}(d_{0k}) = \begin{cases} \dfrac{d - d_{0k}}{\varepsilon_{d_{0k}}} & \text{if} \quad d_{dM_{\min 1}} < d < d_{dM_{\min 2}} \\ 1 & \text{if} \quad d_{dM_{\min 2}} < d < d_{dM_{\max 1}} \\ \dfrac{d_{0k} - d}{\varepsilon_{d_{0k}}} & \text{if} \quad d_{dM_{\max 1}} < d < d_{dM_{\max 2}} \\ 0 & \text{otherwise} \end{cases} \tag{4.16}$$

$$\mu_{dH}(d_{0k}) = \begin{cases} 1 & \text{if} \quad d > d_{dH_{\max}} \\ \dfrac{d - d_{0k}}{\varepsilon_{d_{0k}}} & \text{if} \quad d_{dH_{\min}} < d < d_{dH_{\max}} \\ 0 & \quad \text{otherwise} \end{cases} \tag{4.17}$$

Water volume *v*

The membership functions of $\mu_{vL}(v_{0j})$, $\mu_{vM}(v_{0j})$, $\mu_{vH}(v_{0j})$ corresponding to the water volume v are expressed as follows (Fig. 4.5):

$$\mu_{vL}(v_{0j}) = \begin{cases} 1 & \text{if} \quad 0 < v < v_{vL_{\min}} \\ \dfrac{v_{0j} - v}{\varepsilon_{v_{0j}}} & \text{if} \quad v_{vL_{\min}} < v < v_{vL_{\max}} \\ 0 & \text{otherwise} \end{cases} \tag{4.18}$$

$$\mu_{vM}(v_{0j}) = \begin{cases} \dfrac{v - v_{0j}}{\varepsilon_{v_{0j}}} & \text{if} \quad v_{vM_{\min1}} < v < v_{vM_{\min2}} \\ 1 & \text{if} \quad v_{vM_{\min2}} < v < v_{vM_{\max1}} \\ \dfrac{v_{0j} - v}{\varepsilon_{v_{0j}}} & \text{if} \quad v_{vM_{\max1}} < v < v_{vM_{\max2}} \\ 0 & \text{otherwise} \end{cases} \tag{4.19}$$

$$\mu_{vH}(v_{0j}) = \begin{cases} 1 & \text{if} \quad v > v_{vH_{\max}} \\ \dfrac{v - v_{0j}}{\varepsilon_{v_{0j}}} & \text{if} \quad v_{vH_{\min}} < v < v_{vH_{\max}} \\ 0 & \text{otherwise} \end{cases} \tag{4.20}$$

Power difference Δ*P*

The membership functions of $\mu_{fL}(f_{0e})$, $\mu_{fH}(f_{0e})$ corresponding to ΔP are expressed as follows:

$$\mu_{fL}(f_{0e}) = \begin{cases} 1 & \text{if} \quad 0 < f < f_{fL_{\min}} \\ \dfrac{f_{0e} - f}{\varepsilon_{f_{0e}}} & \text{if} \quad f_{fL_{\min}} < f < f_{fL_{\max}} \\ 0 & \text{otherwise} \end{cases} \tag{4.21}$$

$$\mu_{fH}(f_{0e}) = \begin{cases} 1 & \text{if} \quad f > f_{fH_{\max}} \\ \dfrac{f - f_{0e}}{\varepsilon_{f_{0e}}} & \text{if} \quad f_{fH_{\min}} < f < f_{fH_{\max}} \\ 0 & \text{otherwise} \end{cases} \tag{4.22}$$

Switching control of the relays R_l, R_b, R_{lb}

The relays membership functions $\mu_{\text{off}\ r_l,r_b,r_{lb}}(o_{0z})$, $\mu_{\text{on}\ r_l,r_b,r_{lb}}(o_{0z})$ corresponding to the relays R_l, R_b, and R_{lb} are expressed as follows (Fig. 4.5):

$$\mu_{\text{off}\ r_l,r_b,r_{lb}}(o_{0z}) = \begin{cases} 1 & \text{if} \quad 0 < o < o_{\text{off}_{\min}} \\ \dfrac{o_{0z} - o}{\varepsilon_{o_{0z}}} & \text{if} \quad o_{\text{off}_{\min}} < o < o_{\text{off}_{\max}} \\ 0 & \text{otherwise} \end{cases} \tag{4.23}$$

$$\mu_{\text{on}\ r_l,r_b,r_{lb}}(o_{0z}) = \begin{cases} 1 & \text{if} \quad o > o_{\text{on}_{\max}} \\ \dfrac{o - o_{0z}}{\varepsilon_{o_{0z}}} & \text{if} \quad o_{\text{on}_{\min}} < o < o_{\text{on}_{\max}} \\ 0 & \text{otherwise} \end{cases} \tag{4.24}$$

4.4.3.3 Inference diagram

Based on the fuzzified inputs ($\tilde{P}_{\text{pv}}$, *dod*, and v), the fuzzy rules used for the inference diagram decide the relays switching. During the day, the relays switching decisions are given in Table 4.1. At night, if there is no water extraction or the pumped water volume during the day is sufficient for the irrigation, all the relays would be *off*; otherwise, if the battery bank is charged, the relay R_{lb} would be *on*, to pump the missing water volume needed for irrigation.

The numerical value of the signals control $r_{0l,b,\text{lb}}$ for the three relays is obtained from:

$$r_{0l,b,\text{lb}} = \frac{\int_0^1 r_{\text{on}}\, \mu_{r_{\text{on}}}\, dr_{\text{on}}}{\int_0^1 \mu_{r_{\text{on}}}\, dr_{\text{on}}} \tag{4.25}$$

4.4.3.4 Defuzzification

The control of the three relays is deduced as follows (Fig. 4.5):

$$\text{If} \quad r_{l,b,\text{lb}} \le 0.5, \ \text{then} \ R_{l,b,\text{lb}} \ \text{is off} \tag{4.26}$$

$$\text{If} \quad r_{l,b,\text{lb}} > 0.5, \ \text{then} \ R_{l,b,\text{lb}} \ \text{is on} \tag{4.27}$$

Table 4.1 Fuzzification of the knowledge base

$P_{pv/v}$	*L*	*M*	*H*
• *dod* is *dL*			
vl	η_{lb} is on η_l is on η_b is off	η_{lb} is on η_l is on η_b is off	η_{lb} is off η_l is on η_b is on
vm	η_{lb} is on η_l is on η_b is off	η_{lb} is on η_l is on η_b is off	η_{lb} is off η_l is on η_b is on
vh	η_{lb} is off η_l is off η_b is off	η_{lb} is off η_l is off η_b is off	η_{lb} is off η_l is off η_b is off
• *dod* is *dM*			
vl	η_{lb} is on η_l is on η_b is off	η_{lb} is on η_l is on η_b is off	η_{lb} is off η_l is on η_b is on
vm	η_{lb} is on η_l is on η_b is off	η_{lb} is on η_l is on η_b is off	η_{lb} is off η_l is on η_b is on
vh	η_{lb} is off η_l is off η_b is on	η_{lb} is off η_l is off η_b is on	η_{lb} is off η_l is off η_b is on
• *dod* is *dH*			
vL	η_{lb} is off η_l is off η_b is on	η_{lb} is off η_l is off η_b is on	η_{lb} is off η_l is on η_b is on
vM	η_{lb} is off η_l is off η_b is on	η_{lb} is off η_l is off η_b is on	η_{lb} is off η_l is on η_b is on
vH	η_{lb} is off η_l is off η_b is on	η_{lb} is off η_l is off η_b is on	η_{lb} is off η_l is off η_b is on

4.5 APPLICATION TO A CASE STUDY

In order to test the efficiency of the proposed algorithm, the EMA is tested and validated using data of the target application: A land planted with tomatoes, located in Medjez El Beb, Northern Tunisia, presented previously in

Section 1.1. This application is prompted by the fact that tomatoes must be irrigated regularly, especially during flowering and fruit formation (Shankara, De Jeude, De Goffau, Hilmi, & Van Dam, 2005): The vegetative cycle in the target area is given from *March* to *July*.

The irrigation is gravity-based: 200 m^3/h starting just before sunrise, to irrigate a 10 ha field by a low-pressure, gravity-driven drip system. For this, a 4.5 kW pump submerged in an 80 m well and a 1800 m^3 reservoir are available. First, the PV and batteries were sized using the algorithm presented in Section 3.3 (Yahyaoui, Chaabene, & Tadeo, 2013): A 10.74 kW PV system (101.5 m^2 panel surface equipped with an MPP Tracker) and a battery bank composed of 8 batteries (210 Ah/12 V) with regulator have been selected. Using the algorithm presented in Section 4.4, the energy management is described now in detail (Fig. 4.4).

4.5.1 Algorithm parameterization

4.5.1.1 PV power P_{pv}

The PVPs nonlinear model, detailed in Section 2.3, is used here to evaluate the PV power P_{pv} (Campana et al., 2013). Thus, P_{pv} is classified as follows:

a. If $P_{pv} \in [0 \;\; 10 \text{ W}]$, then P_{pv} is considered *low.*

b. If $P_{pv} \in [10 \text{ W} \;\; 4500 \text{ W}]$, then P_{pv} is considered *medium*.

c. If $P_{pv} \in [4500 \text{ W} \;\; 10{,}740 \text{ W}]$, then P_{pv} is considered *high*.

4.5.1.2 Battery dod

The battery nonlinear model (Sallem et al., 2009a, 2009b; Serna & Tadeo, 2014), detailed in Section 2.3, is used here to evaluate the *dod*, which is classified as follows:

i. If $dod \in [0 \;\; 0.02]$, then *dod* is considered *low.*

ii. If $dod \in [0.02 \;\; 0.8]$, then *dod* is considered *medium*.

iii. If $dod \in [0.8 \;\; 1]$, then *dod* is considered *high*.

4.5.1.3 Stored water volume v

Using the water demand model (Hahn, 2011; Shankara et al., 2005) (Section 3.3), the water volume v corresponding to each month of the vegetative cycle of tomatoes at the target location is:

1. The mean water volume of *March* (m_1) is $l_1 = 60$ m^3/day.
2. The mean water volume of *April* (m_2) is $l_2 = 100$ m^3/day.
3. The mean water volume of *May* (m_3) is $l_3 = 179$ m^3/day.
4. The mean water volume of *June* (m_4) is $l_4 = 241$ m^3/day.
5. The mean water volume of *July* (m_5) is $l_5 = 321$ m^3/day.

Table 4.2 Water fuzzification corresponding to each month *M*

Month M	m_1	m_2	m_3	m_4	m_5
L					
l_1	Low	Low	Low	Low	Low
l_2	High	Medium	Low	Low	Low
l_3	High	High	Medium	Low	Low
l_4	High	High	High	Medium	Low
l_5	High	High	High	High	Medium

The fuzzification of the water volume depends on the month $_{\text{and}}$ is described in Table 4.2.

4.5.2 Results and discussions

The management algorithm was implemented using the models presented by Yahyaoui et al. (2014a, 2014b). Simulations were carried out using data (solar irradiation, ambient temperature, rainfall, etc.) from the target location (Medjez El Beb) for the irrigation season from *March* to *July*. Obtained results (Figs. 4.6–4.11) prove that the algorithm fulfills the objectives: Relays switching ensures the system autonomy. The water demand is fulfilled and the battery and load are correctly disconnected when not used.

In fact, on *March* 13th ($f_i = 3$), the pump is supplied by the panel and/or the battery, following the algorithm constraints and goals (Fig. 4.6). Indeed, at night, the tank contains the water needed by the crops (60 m^3). So the irrigation step is finished 1 h before sunrise to allow a better absorption of the water by the crops (Saleh, Ozawa, & Khondaker, 2007). Thus, during irrigation (Δ_{irg}), the water volume decreases, following the constant irrigation flow rate (200 m^3/h). At 8:45 am, the tank is empty and the battery is charged. Hence, the system uses both energies from the panel and the battery to pump water. At 4:00 pm, the tank is full. Hence, the available PV energy is used to charge the battery since the battery is not full charged, so only R_b is switched *on*. During all of these modes, the *dod* is always maintained between the prefixed values (0.02 and 0.8), which guarantees the battery safe operation.

Then, on *March* 14th, the energy generated is used to full the reservoir with water, since there is no irrigation. In *March* 15th, the PV energy is not used since the reservoir and the battery bank are full.

On *March* 16th (Fig. 4.6), from the starting defect time $t_{sd} = 3{:}50$ am until the end defect time $t_{ed} = 5{:}00$ am, an additional water extraction is applied with a flow rate = 150 m^3/h. Thus, the developed algorithm allows the relay R_{lb} to be switched *on*, which connects the battery bank to the pump, so as to compensate the loss in

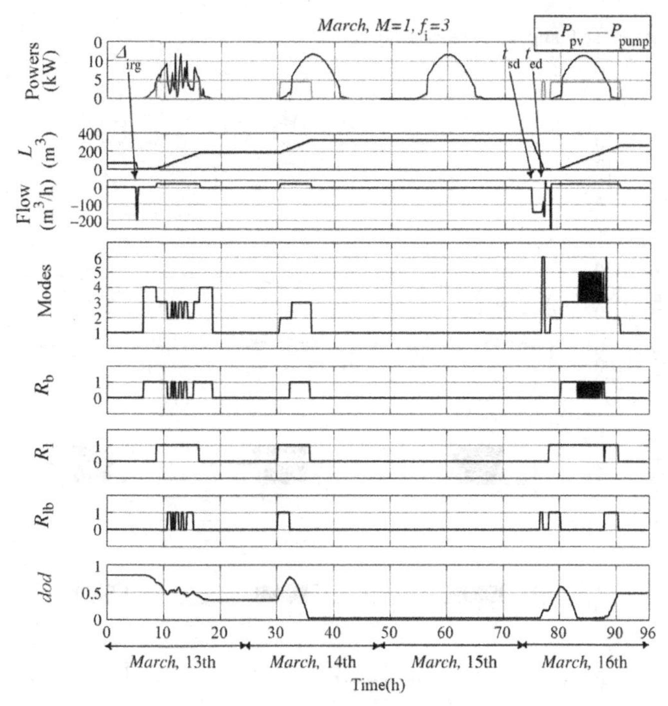

Figure 4.6 Algorithm response in the case study for 4 days in *March* ($f_i = 3$).

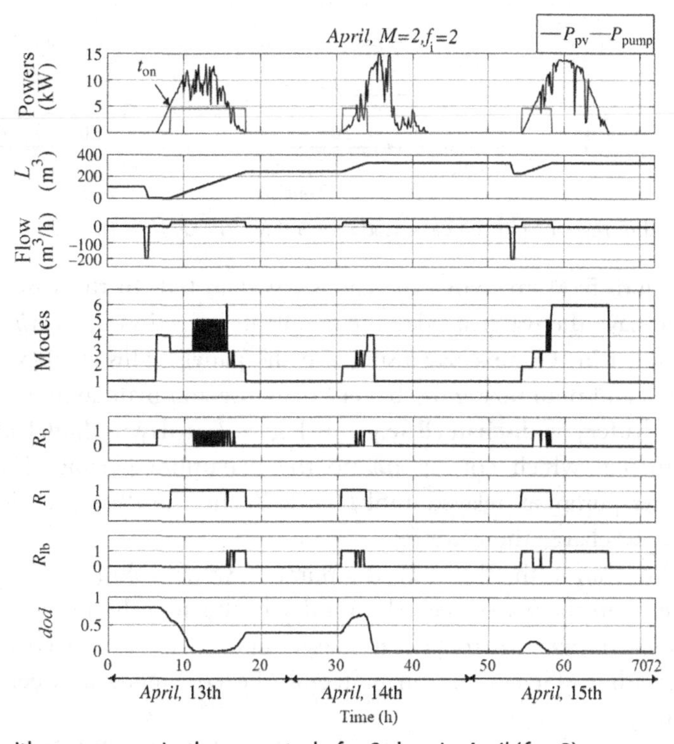

Figure 4.7 Algorithm response in the case study for 3 days in *April* ($f_i = 2$).

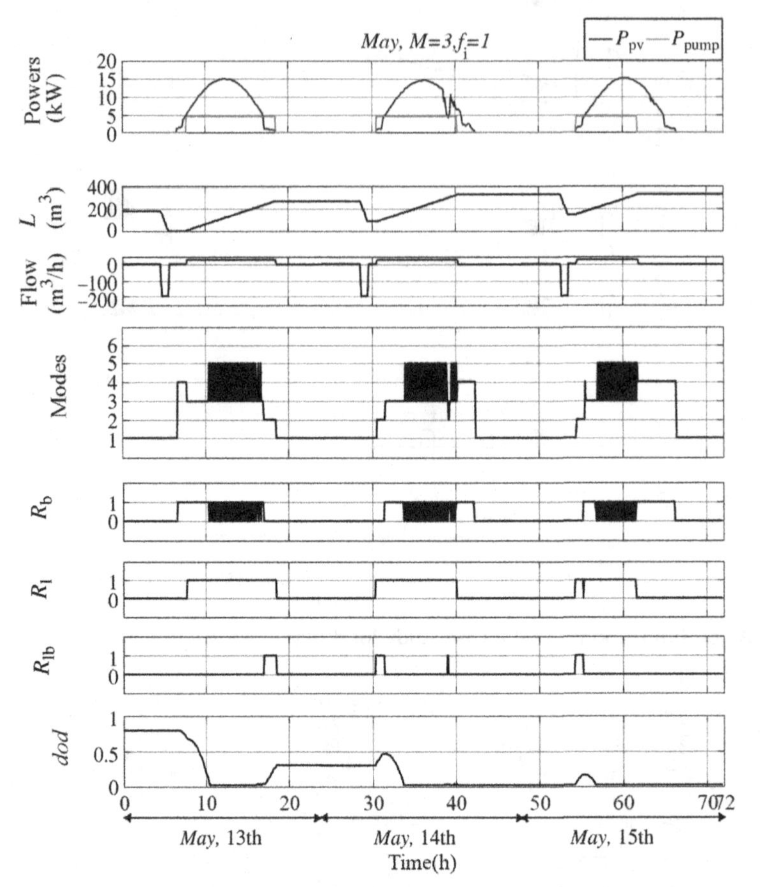

Figure 4.8 Algorithm response in the case study for 3 days in *May* ($f_i = 1$).

water volume until 5:00 am, while the *dod* is less than 0.8. In this case, the pumping is performed to have the water needed for the plants irrigation in *March*, since the reservoir can be filled in the next day, using also the panels. This strategy allows having the reservoir full and minimizing the battery bank use and maintaining it full charged.

In case the battery is discharged (e.g., in Fig. 4.7 at dawn *April* 13th), the initial *dod* is equal to 0.8, which corresponds to the maximum permitted value for *dod*. Hence, the water pumping delayed until t_{on} = 8:20 am; the PV energy is then temporarily used only to charge the batteries.

At the end of *May* 13th (Fig. 4.8), the pumped water volume is equal to 227 m^3, which is higher than the water needed for a day in this month (179 m^3). For this reason, the pumping process is stopped at sunset time t_{ss}, to save electric energy. This proves the algorithm efficiency in guaranteeing the coherence between saving water and electric energy.

To clarify the internal working of the algorithm, the control defuzzification signals of relays are presented for specific days of *June* and *July*, selected as there were rapid

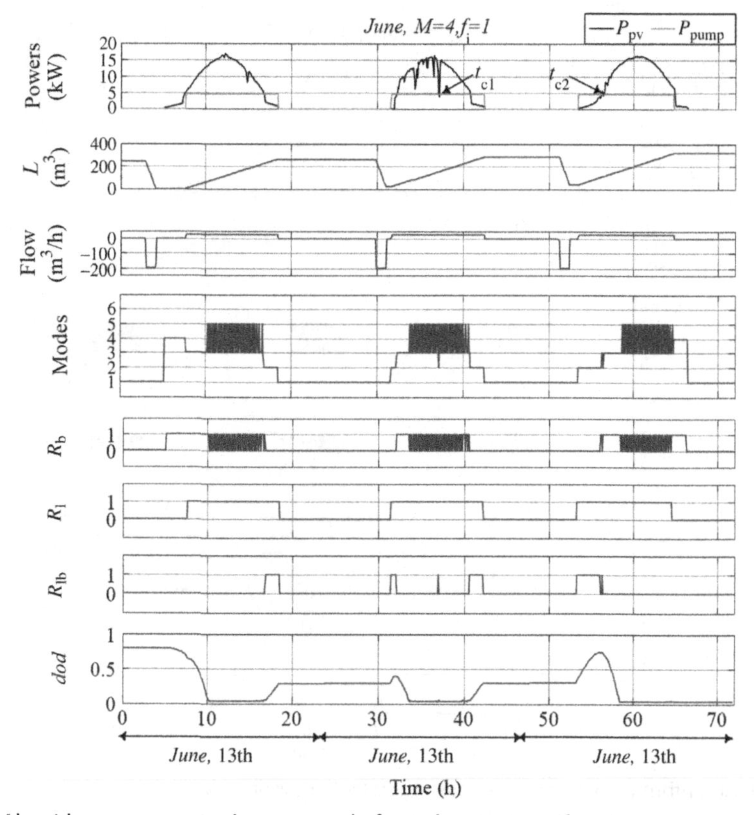

Figure 4.9 Algorithm response in the case study for 3 days in *June* ($f_i = 1$).

changes in the PV power, for example at t_{c_1}, t_{c_2}, t_{c_3}, and t_{c_4} (Figs. 4.9 and 4.10). It can be seen that the control signals ensure relays complementary switching (relays R_b and R_{lb}) since each relay is considered *on* when the membership degree for the relay control signal is higher than 0.5 otherwise it is *off*, enabling then a continuous power supply for the pump and the system autonomy. This is detailed in Fig. 4.11 where the FMA ensures pumping the water volume expected for *July* and the *dod* to be less than 0.8.

The proposed algorithm is evaluated from *March* to *July* as this is the growing season for tomatoes in the target location. It is clear that it ensures pumping more water volume than needed, especially during *March* and *April* (Fig. 4.12), since the water volume pumped L is significantly higher than the water volume needed for the crops irrigation. Moreover, the use average of batteries is minimized: the battery bank maximum contribution in supplying the pump represents 26% of the panels contribution (Fig. 4.13). This proves the algorithm efficiency in keeping the battery bank charged and minimizing its use.

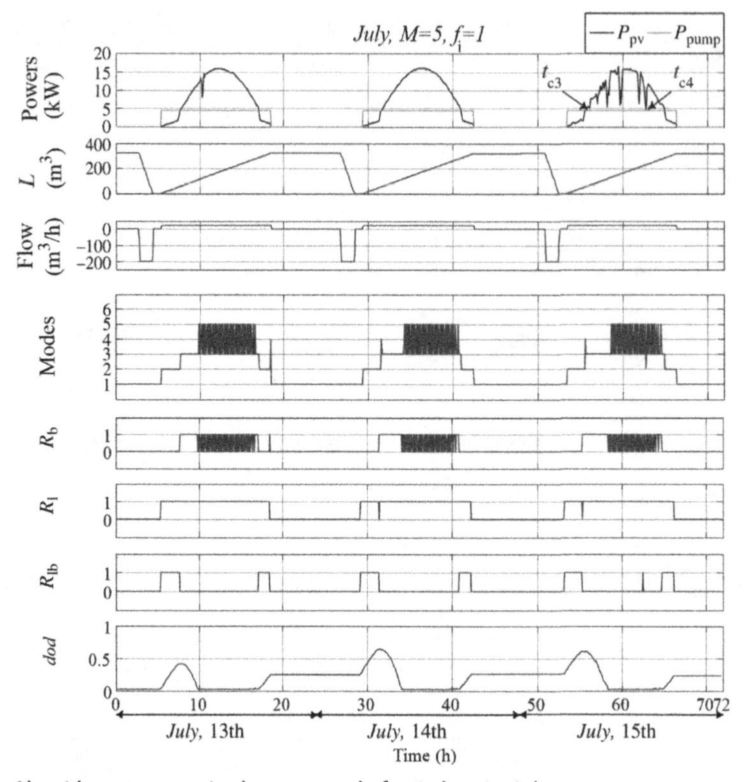

Figure 4.10 Algorithm response in the case study for 3 days in *July*.

4.6 EXPERIMENTAL VALIDATION

In this section, the management algorithm proposed in Section 4.4 is validated. For this, an installation was designed and installed in the laboratory of Automatic Control of the School of Industrial Engineering, University of Valladolid, Spain. The plant is composed of a programmable power supply (PPS), which generates the PV power, a lead–acid battery 12 Ah/12 V, an inverter, a pump, two reservoirs, sensors of pressure and current, and an acquisition card, installed in a computer, as it is shown in Fig. 4.14.

4.6.1 Installation description

The installation allows the PV power to be generated, sensors data to be acquired and recorded in the computer, and the relays control signals to be generated by the control algorithm presented in Section 4.4. These functions are described below (Fig. 4.14).

- *PV power generation*: This step is ensured via the PPS which, in the studied plant, substitutes the PVPs and the chopper (that tracks the MPP), in order to make the results reproducible, making possible to evaluate the control algorithm in different

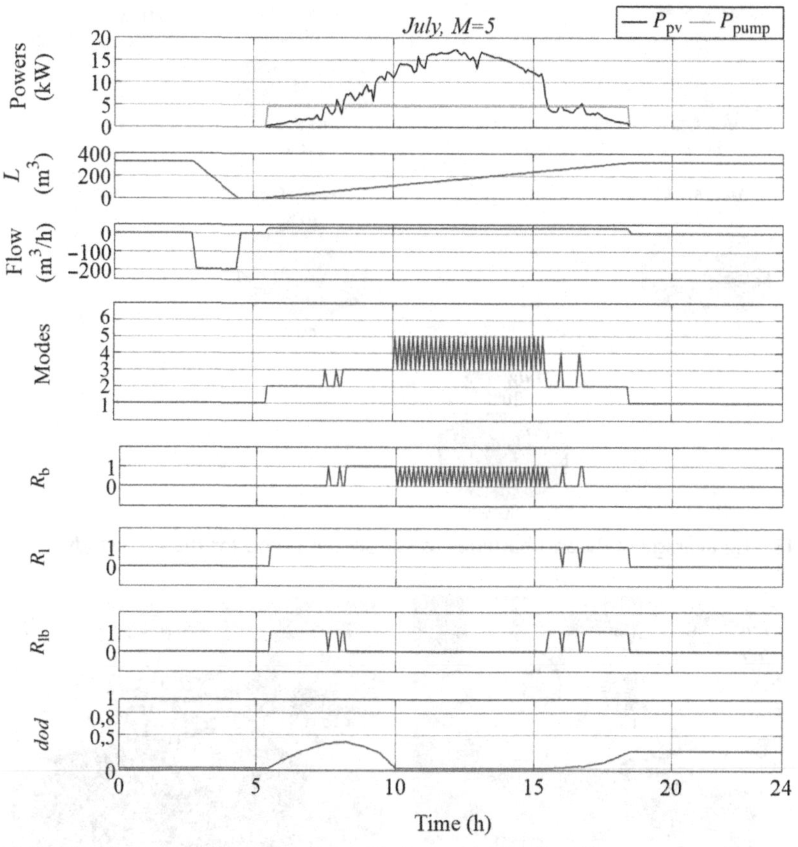

Figure 4.11 Defuzzification of the control signals for a specific day of *July*.

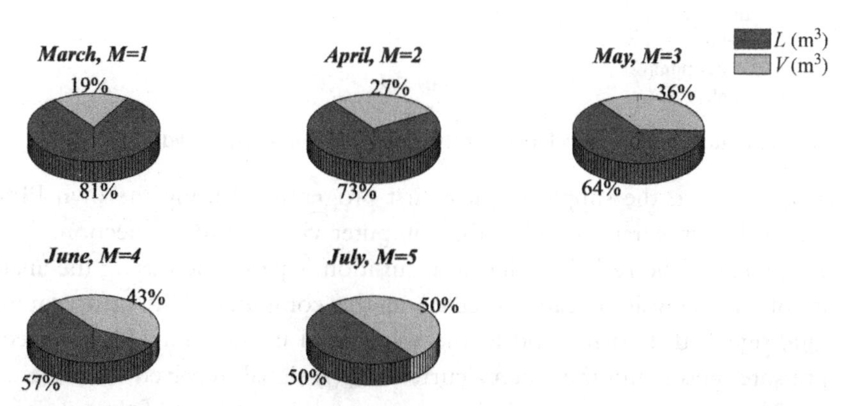

Figure 4.12 Needed and possible pumped water volume rates during the tomato vegetative cycle.

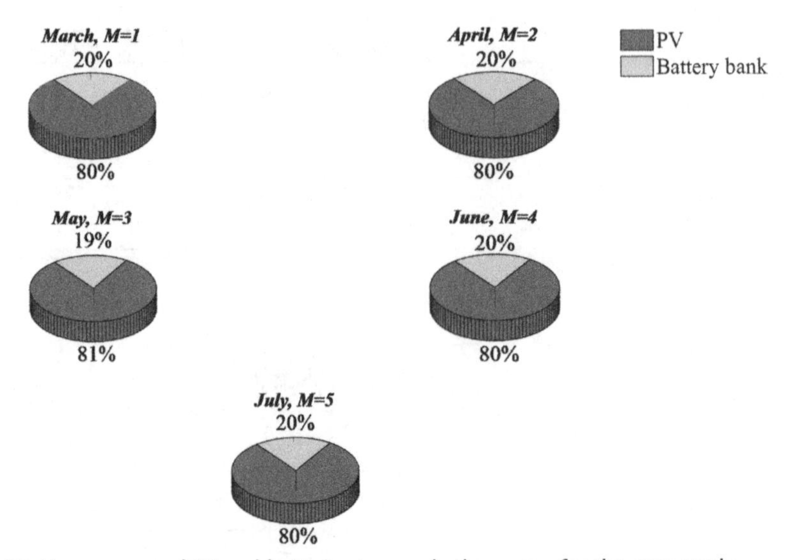

Figure 4.13 Use average of PV and batteries to supply the pump for the case study.

Figure 4.14 Laboratory system developed for the energy management validation.

situations. In fact, the supply signal is first programmed using the own PPS software, and then it is transmitted to the computer via an USB connection.

- *Data acquisition*: The real-time signals acquisition is performed using the analogical inputs of the acquisition card, installed in the computer. This card acquires the analogic signals that correspond to the water level in the tank $L(t)$, received from the pressure sensor, and the battery current $I_{bat}(t)$ signal, received from the current sensor. These signals are then used to generate the switching of the relays control signals, using the algorithm proposed in Section 4.4.

- *Generation of the relays control signals*: The acquired sensors signals are used by the algorithm developed in Section 4.4, and implemented in Matlab, in order to generate the control signals for the relays R_l, R_b, and R_{lb} via the outputs of the acquisition card.

4.6.2 Cases study validation

In this section, some cases study proposed in Section 4.5 are validated (Figs. 4.15–4.19). The management algorithm proposed is tested for typical days for each month during the crop vegetative cycle. This allows validating the possible operating modes discussed in Section 4.4.2. Fig. 4.15 shows that in *March*, when the panel generates power less than the demanded by the pump and the reservoir is not full, the

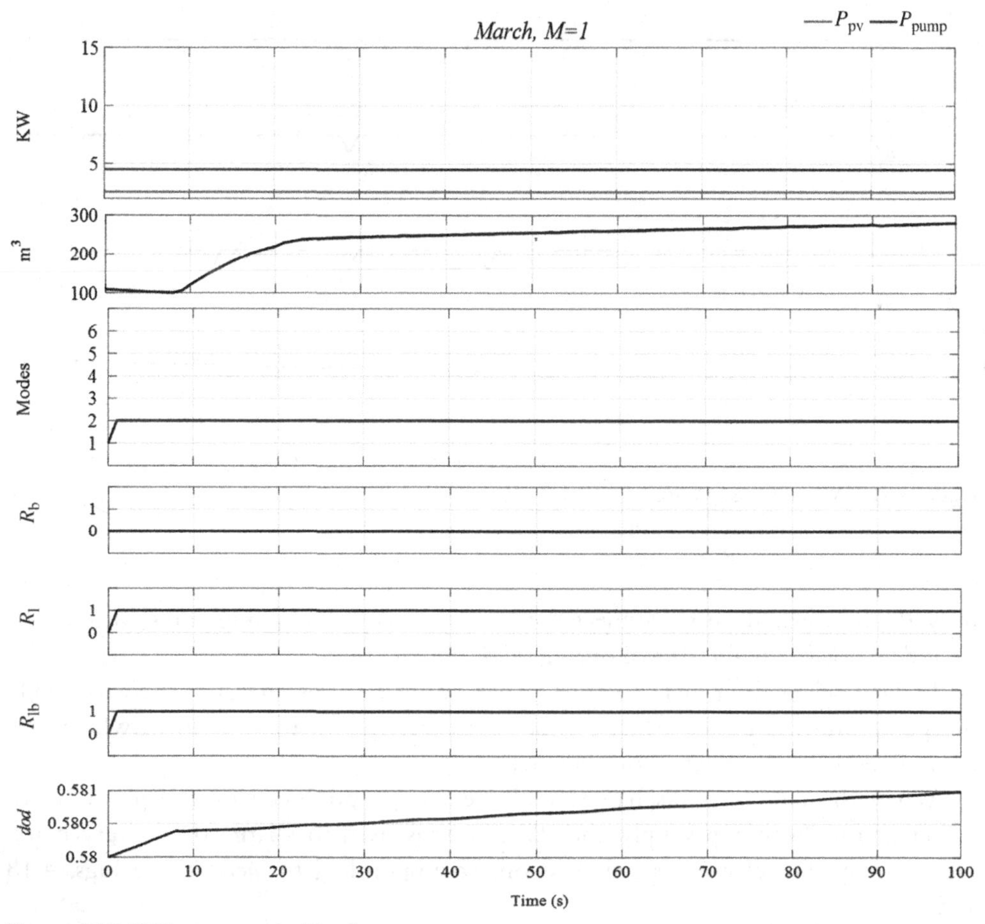

Figure 4.15 EMA response in *March*.

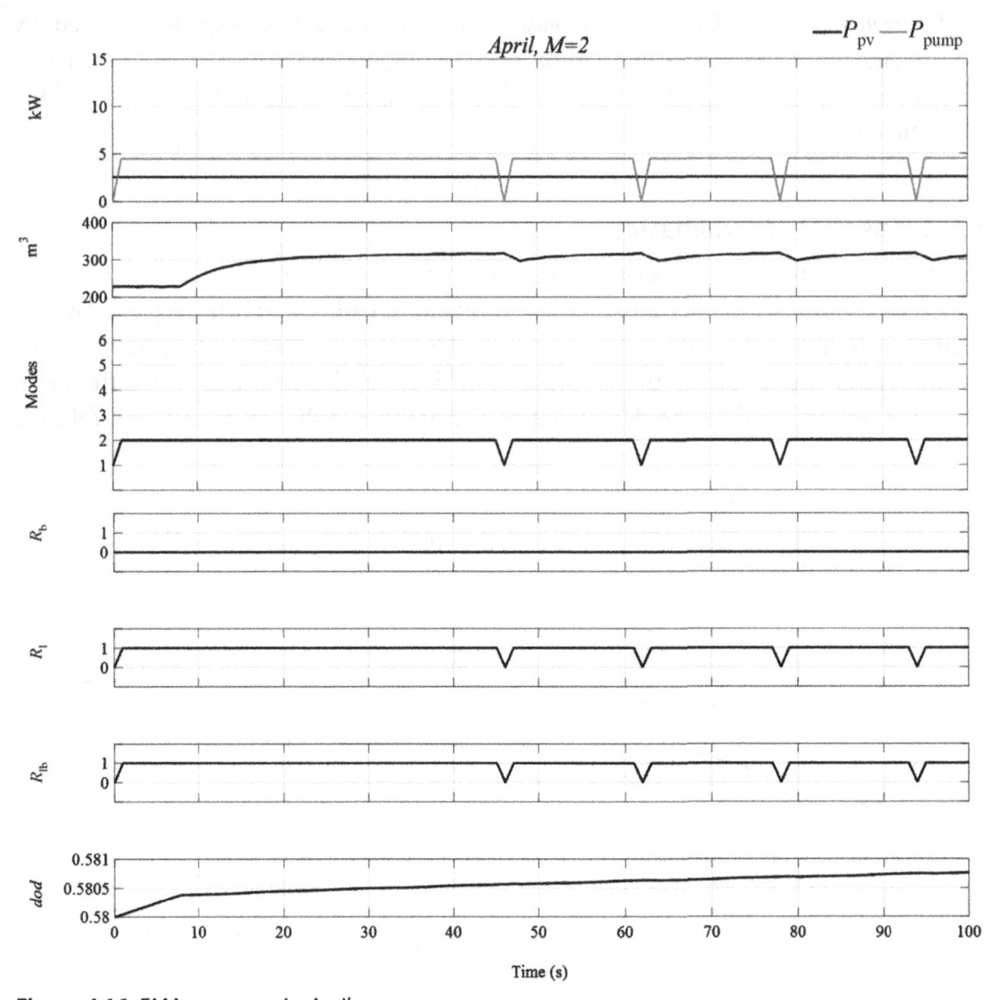

Figure 4.16 EMA response in *April.*

relay R_{lb} is switched on to connect the battery bank to the pump. This corresponds to *mode 2*. *Mode 2* is also shown during *April* (Fig. 4.16).

In *May*, when the panel produces power in excess, this energy is used to supply the pump and charge the battery bank. This corresponds to *mode 3*, in which, the relays R_l and R_b are switched on (Fig. 4.17).

When the battery is discharged and the energy produced by the panel is not sufficient for the pump supply, the PV power is used to charge the battery bank. Hence, only the relay R_b is switched on, corresponding to *mode 4* (see Figs. 4.18 and 4.19).

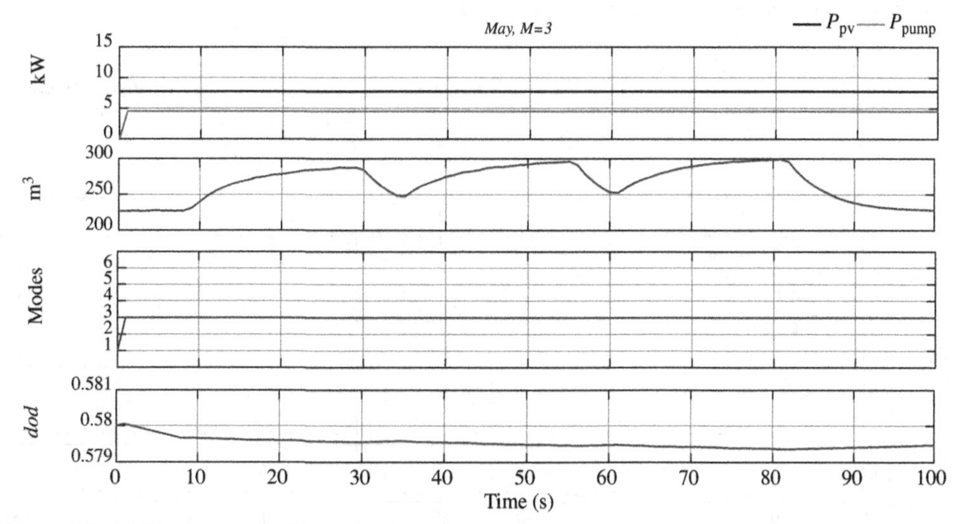

Figure 4.17 EMA response in *May*.

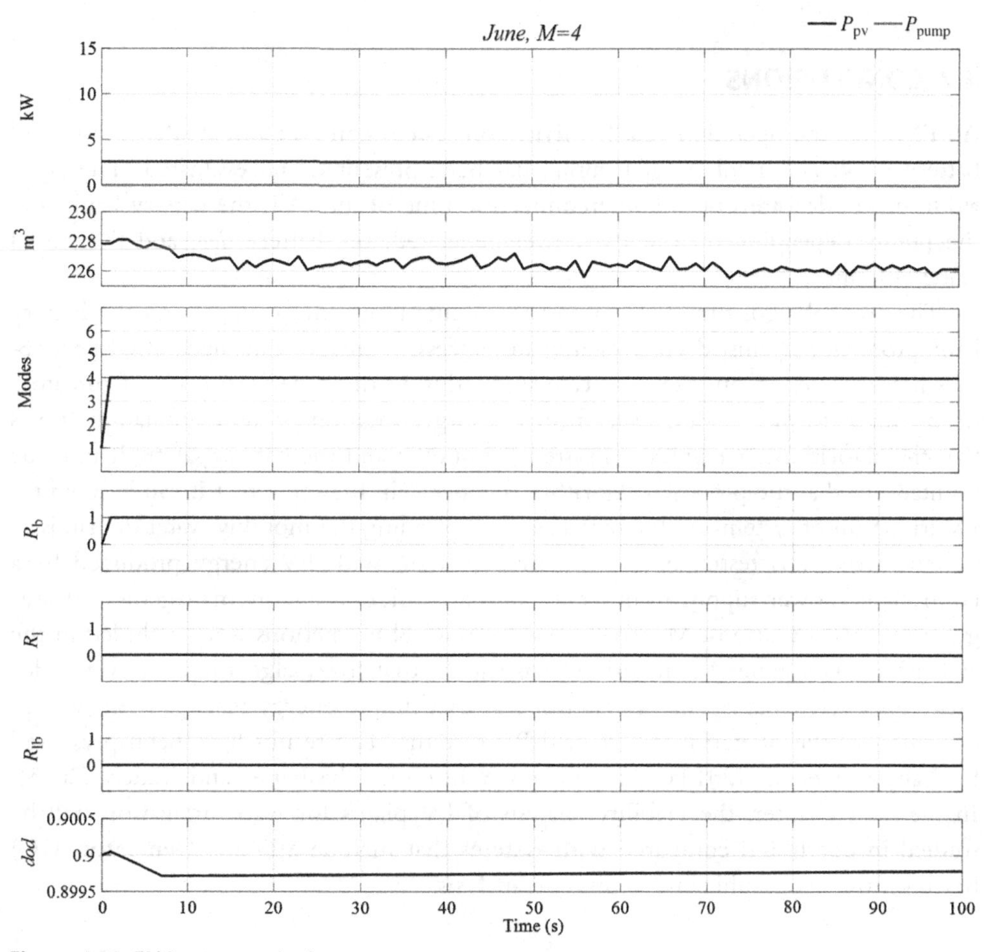

Figure 4.18 EMA response in *June*.

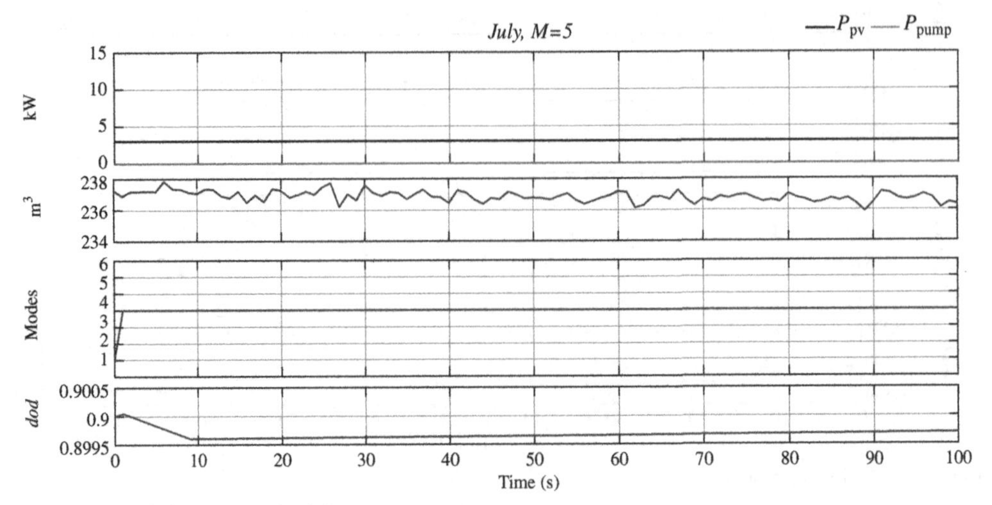

Figure 4.19 EMA response in *July*.

4.7 CONCLUSIONS

An FMA for the operation of a PV irrigation system composed of PVPs, a lead–acid battery bank, and a submerged pump, has been presented and evaluated. The algorithm makes decisions on the interconnection time of the PVP, the battery bank, and the pump, depending on the PV power generated, the battery *dod*, and the stored water amount.

The control algorithm aims to ensure a continuous pump supply and the battery bank protection against deep discharge and excessive charge. The algorithm effectiveness has been tested on a specific case study, during the vegetative cycle of tomatoes (from *March* to *July*). Using data from the target location, system simulation shows that the algorithm guarantees the system autonomy and the battery safety. It must be pointed out that the proposed algorithm is general, in the sense that it can be used for PV irrigation of systems of different sizes, by providing the monthly water demands.

The work was tested in a 1:150 pilot system, with PV energy produced by a controllable power supply, in order to compare different energy management strategies in similar situations. Moreover, meteorological predictions were included in the algorithm. As a general conclusion, it is shown that fuzzy algorithms are very adequate for energy management systems, and that simple energy management systems can improve the operation of off-grid PV systems. The results have been presented by Yahyaoui et al. (2014a, 2014b) and Yahyaoui, Chaabene, and Tadeo (2015). In the next chapter, the viability analysis of PV plants for water irrigation will be studied in depth and compared with systems that include a Diesel Generator (DG) based on the geographic and economic analysis.

REFERENCES

Abdeen, M. O. (2008). Energy, environment and sustainable development. *Renewable and Sustainable Energy Reviews, 12*(9), 2265–2300.

Al-Alawi, A., Al-Alawi, S. M., & Islam, S. M. (2007). Predictive control of an integrated PV-diesel water and power supply system using an artificial neural network. *Renewable Energy, 32*(8), 1426–1439.

Arrouf, M., & Ghabrour, S. (2007). Modelling and simulation of a pumping system fed by photovoltaic generator within the Matlab/Simulink programming environment. *Desalination, 209*(1), 23–30.

Beccali, M., Cellura, M., & Ardente, D. (1998). Decision making in energy planning: The ELECTRE multicriteria analysis approach compared to a fuzzy-sets methodology. *Energy Conversion and Management, 39*(16), 1869–1881.

Belgacem, B. G. (2012). Performance of submersible PV water pumping systems in Tunisia. *Energy for Sustainable Development, 16*(4), 415–420.

Ben Salah, C., Chaabene, M., & Ben Ammar, M. (2008). Multi-criteria fuzzy algorithm for energy management of a domestic photovoltaic panel. *Renewable Energy, 33*(5), 993–1001.

Berrazouane, S., & Mohammedi, K. (2014). Parameter optimization via cuckoo optimization algorithm of fuzzy controller for energy management of a hybrid power system. *Energy Conversion and Management, 78*, 652–660.

Betka, A., & Attali, A. (2010). Optimization of a photovoltaic pumping system based on the optimal control theory. *Solar Energy, 84*(7), 1273–1283.

Bouchon-Meunier, B., Foulloy, L., & Ramdani, M. (1994). *La logique floue* (Vol. 2702). Presses universitaires de France.

Campana, P. E., Li, H., & Yan, J. (2013). Dynamic modelling of a PV pumping system with special consideration on water demand. *Applied Energy, 112*, 635–645.

Celik, A. N., & Acikgoz, N. (2007). Modelling and experimental verification of the operating current of mono-crystalline photovoltaic modules using four- and five-parameter models. *Applied Energy, 84*(1), 1–15.

Cheikh, M. A., Aïssa, B. H., Malek, A., & Becherif, M. (2010). Mise au point d'une régulation floue pour serre agricole à énergie solaire. *Revue des Energies Renouvelables, 13*(3), 421–443.

Etienne, S., Alberto, T., & Mikhail, S. (2011). Explicit model of photovoltaic panels to determine voltages and currents at the maximum power point. *Solar Energy, 85*(5), 713–722.

Gergaud, O. (2002). Modélisation énergétique et optimisation économique d'un système éolien et photovoltaïque couplé au réseau et associé à un accumulateur (Thesis). France: Normal School of Cachan.

Ghoneim, A. A. (2006). Design optimization of photovoltaic powered water pumping systems. *Energy Conversion and Management, 47*(11), 1449–1463.

Hahn, F. (2011). Fuzzy controller decreases tomato cracking in greenhouses. *Computers and Electronics in Agriculture, 77*(1), 21–27.

Hemanth, D. J., Vijila, C. K. S., & Anitha, J. (2010). Application of neuro-fuzzy model for MR brain tumor image classification. *Biomedical Soft Computing and Human Sciences, 16*(1), 95–102.

Jung, J. H., & Ahmed, S. (2012). Real-time simulation model development of single crystalline photovoltaic panels using fast computation methods. *Solar Energy, 86*(6), 1826–1837.

Kahraman, C., Kaya, I., & Cebi, S. (2009). A comparative analysis for multiattribute selection among renewable energy alternatives using fuzzy axiomatic design and fuzzy analytic hierarchy process. *Energy, 34*(10), 1603–1616.

Kalogirou, S. A. (2001). Artificial neural networks in renewable energy systems applications: A review. *Renewable and Sustainable Energy Reviews, 5*(4), 373–401.

Lee, J. C., Lin, W. M., Liao, G. C., & Tsao, T. P. (2011). Quantum genetic algorithm for dynamic economic dispatch with valve-point effects and including wind power system. *International Journal of Electrical Power & Energy Systems, 33*(2), 189–197.

Liu, X., Fang, X., Qin, Z., Ye, C., & Xie, M. (2011). A short-term forecasting algorithm for network traffic based on chaos theory and SVM. *Journal of Network and Systems Management, 19*(4), 427–447.

Lo Brano, V., Orioli, A., Giuseppina, C., & Di Gangi, A. (2010). An improved five-parameter model for photovoltaic modules. *Solar Energy Materials and Solar Cells, 94*(8), 1358–1370.

Messikh, L., Chikhi, S., Chikhi, F., & Chergui, T. (2008). Mise au point d'un régulateur de charge/décharge de batterie avec seuils adaptatifs de tension pour les applications photovoltaïques. *Revue des Energies Renouvelables*, *11*(2), 281–290.

Moradi, M. H., & Abedini, M. (2012). A combination of genetic algorithm and particle swarm optimization for optimal DG location and sizing in distribution systems. *International Journal of Electrical Power & Energy Systems*, *34*(1), 66–74.

Moradi-Jalal, M., Marino, M. A., & Afshar, A. (2003). Optimal design and operation of irrigation pumping stations. *Journal of Irrigation and Drainage Engineering*, *129*(3), 149–154.

Saad, N., & Arrofiq, M. (2012). A PLC-based modified-fuzzy controller for PWM-driven induction motor drive with constant V/Hz ratio control. *Robotics and Computer-Integrated Manufacturing*, *28*(2), 95–112.

Saleh, M. I., Ozawa, K., & Khondaker, N. A. (2007). Effect of irrigation frequency and timing on tomato yield, soil water dynamics and water use efficiency under drip irrigation. *Proceedings of the eleventh international water technology conference (IWTC)* (Vol. 1, pp. 15–18).

Sallem, S., Chaabene, M., & Kamoun, M. B. A. (2009a). Energy management algorithm for an optimum control of a photovoltaic water pumping system. *Applied Energy*, *86*(12), 2671–2680.

Sallem, S., Chaabene, M., & Kamoun, M. B. A. (2009b). Optimum energy management of a photovoltaic water pumping system. *Energy Conversion and Management*, *50*(11), 2728–2731.

Semaoui, S., Hadj Arab, A., Bacha, S., & Azoui, B. (2013). The new strategy of energy management for a photovoltaic system without extra intended for remote-housing. *Solar Energy*, *94*, 71–85.

Serna, Á., & Tadeo, F. (2014). Offshore hydrogen production from wave energy. *International Journal of Hydrogen Energy*, *39*(3), 1549–1557.

Shankara, N., De Jeude, J. V. L., De Goffau, M., Hilmi, M., & Van Dam, B. (2005). Production, processing and marketing. Agromisa Foundation and CTA.

Singh, G. K. (2013). Solar power generation by PV (photovoltaic) technology: A review. *Energy*, *53*, 1–13.

Welch, R. L., & Venayagamoorthy, G. K. (2010). Energy dispatch fuzzy controller for a grid-independent photovoltaic system. *Energy Conversion and Management*, *51*(5), 928–937.

Yahyaoui, I., Chaabene, M., & Tadeo, F. (2013). An algorithm for sizing photovoltaic pumping systems for tomatoes irrigation. *Proceedings of the IEEE international conference on renewable energy research and applications (ICRERA)* (pp. 1089–1095).

Yahyaoui, I., Chaabene, M., & Tadeo, F. (2015). Energy management for photovoltaic irrigation with a battery bank. *International Journal of Energy Optimization and Engineering (IJEOE)*, *4*(3), 18–32.

Yahyaoui, I., Chaabene, M., & Tadeo, F. (2014b). A fuzzy based energy management for a photovoltaic pumping plant for tomatoes irrigation. *Proceedings of the IEEE international renewable energy congress (IREC)* (pp. 122–127).

Yahyaoui, I., Chaabene, M., & Tadeo, F. (2015). Fuzzy energy management for photovoltaic water pumping system. *International Journal of Computer Applications (IJCA)*, *110*(9), 29–36.

Yahyaoui, I., Sallem, S., Kamoun, M. B. A., & Tadeo, F. (2014). A proposal for management of off-grid photovoltaic systems with non-controllable loads using fuzzy logic. *Energy Conversion and Management*, *78*, 835–842.

CHAPTER 5

Viability of DG, DG/PV and PV/Batteries Plants for Water Pumping: Sensitivity Analysis According to Geographical and Economic Parameters Variations

5.1 INTRODUCTION

As it has been seen in the previous chapters, in remote agriculture areas, a PV water pumping system equipped with a battery bank is a promoting solution, thanks to several advantages; namely, its continuous decreasing price, the easiness in installing near to the place of consumption, and the nonemission of greenhouse gases. Hence, studied from an energetic point of view, this type of installations can fulfill the pump demands, by providing it the needed electrical power, and ensure a safe operating for the system components, if it is optimally designed and managed. However, in the same areas, since agriculture farms require electrical energy for several applications including water pumping, farm operation, lighting, etc., diesel generators (DG) are also used to produce electricity (Al-Alawi, Al-Alawi, & Islam, 2007).

Actually, there are many issues related to the use of DG for the electricity generation, such as the need for a continuous fuel providing and therefore, the necessity of the user presence, the control complexity, due to the starting time delays, especially when used as a backup with other energy sources (Al-Alawi et al., 2007; Wichert, 1997). These issues are not studied here. Indeed, the aim of this chapter is to compare the cost of water pumping systems supplied by three types of sources: first, a DG supplying the pump. The second system type is supplied by PVPs and a DG. Finally, the third system consists of a PV water pumping system equipped with a battery bank (which has been detailed in the previous chapters). Then, the sensitivity analysis of PV water pumping systems is studied, based on the geographic, and economic variations. Hence, this chapter is organized as follows: Section 5.2 is concerned with the study of water pumping systems types, which include DGs. In this section, DG systems for water pumping are detailed, and a model for the genset is studied in depth. The economic viability of the mentioned water pump plants is studied in Section 5.3, using climatic and economic

Specifications of Photovoltaic Pumping Systems in Agriculture
DOI: http://dx.doi.org/10.1016/B978-0-12-812039-2.00005-3

data of Tunisia. Then, in Section 5.4, the costs of the three water pumping systems are analyzed and compared in three different countries, which are Tunisia, Spain, and Qatar, based on climatic, geographic, and economic parameters of these countries. The two countries (Spain and Qatar) are chosen based on the similarity of oil prices, and since they are characterized by important and similar amounts of solar energy, compared with Tunisia. Finally, Section 5.5 presents the chapter conclusion.

5.2 WATER PUMPING SYSTEMS EQUIPPED WITH DG

5.2.1 DG systems for water pumping systems

In this case, the water pump is supplied by the DG, which is the unique power source in such systems. These systems are generally used in areas where the price of fuel is cheap and/or in countries where renewable energies use still not common or expensive (Fig. 5.1).

5.2.2 PV/DG systems

Some other systems composed of PV—DG can also be used for agriculture applications. These systems do not include storage devices for energetic or economic reasons. Indeed, in such systems, the DG operates when the power generated by the PVPs is not sufficient to supply the water pump or during the night (Wichert, 1997) (Fig. 5.2).

5.2.3 PV/batteries/DG systems

In addition to PV/battery/pump systems, which have been studied in the previous chapters, there exist hybrid systems composed of PV panels, a DG and, in some cases, equipped with small size battery storage bank (Fig. 5.3). In fact, hybrid systems equipped with battery banks provide sufficient storage to the load power supply in case of demand, and also ensure that the genset have not undesirable

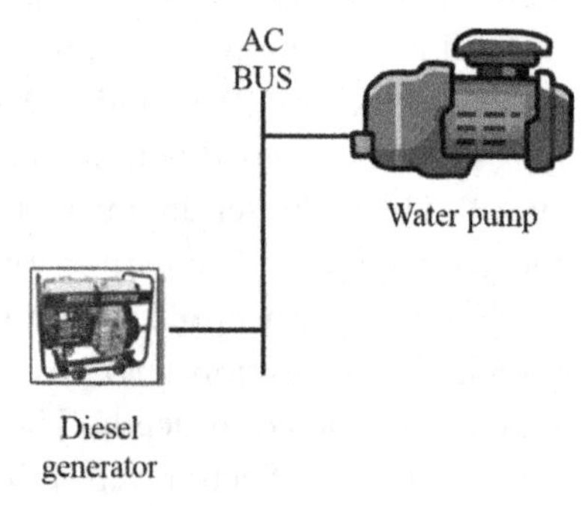

Figure 5.1 Architecture of DG water pumping plants.

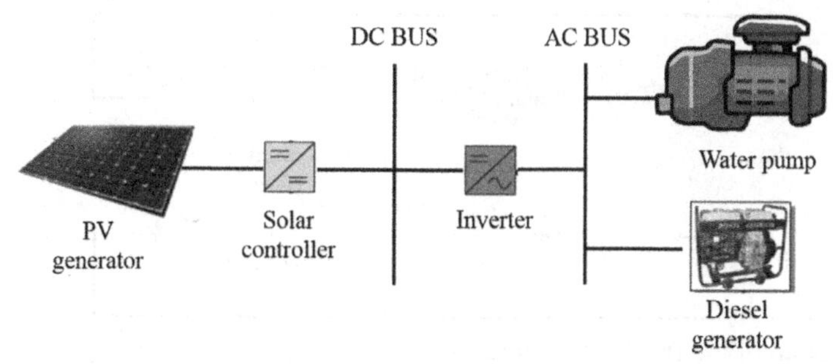

Figure 5.2 Architecture of PV–DG water pumping plants.

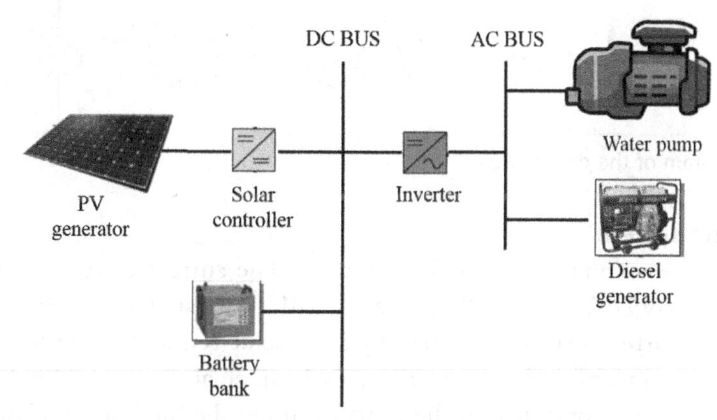

Figure 5.3 PV/batteries/DG systems for water pumping application.

and successive on/off states change, taking into account their starting time-delay (Al-Alawi et al., 2007). Hence, the genset is generally used as a supplement to the battery and the PV panels, since it acts as a back-up generator while the power generated from the PV/batteries is insufficient for the pump supply. Generally, these systems (PV/batteries/ DG) are used in applications that operate during the day and also nights (Bonanno, Consoli, Raciti, Morgana, & Nocera, 1999; Wichert, 1997).

5.2.4 DG system

5.2.4.1 DG modeling

Generally, the DG is composed of three main components: the diesel engine, which includes the mechanical part of the genset, the synchronous generator, in which the mechanical power is converted into electric power, and the excitation system, which is used for the synchronous machine excitation (Fig. 5.4) (Salazar, Tadeo, De Prada, & Palacin, 2013). Each component is detailed in depth below.

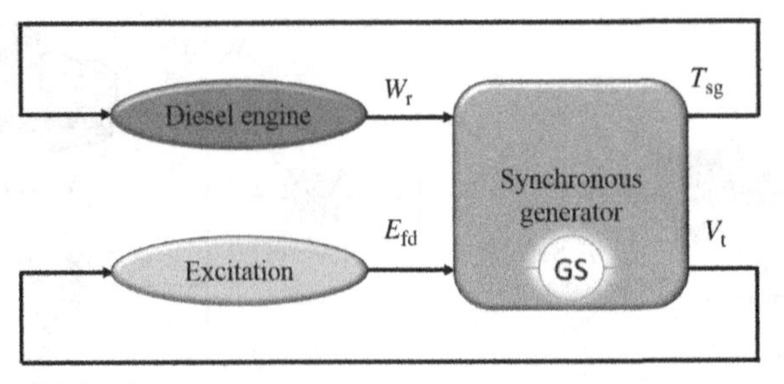

Figure 5.4 Diagram of the DG components.

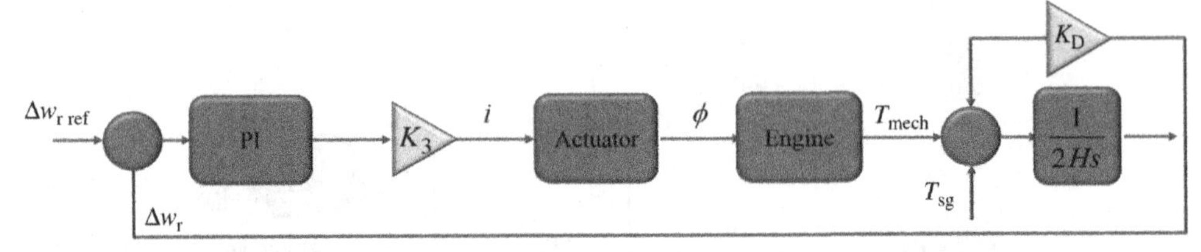

Figure 5.5 Diagram of the diesel engine.

Diesel engine system

The diesel engine is composed of four main parts: The current driver, the actuator, the engine, and the flywheel (Fig. 5.5) (Salazar et al., 2013; Salazar, Tadeo, & Prada, 2015).

In fact, the current driver, described by a constant K_3, allows the fuel flow to be converted into a mechanical torque T_{mech} (Salazar et al., 2015). Then, the engine speed is automatically controlled in the actuator, using the fuel intake. Hence, the fuel amount ϕ injected into the combustion chamber controls the current i (Graciano, Vargas, & Ordonez, 2015). Indeed, the injected fuel is ignited by the compressed hot air in the combustion chamber located at the "engine" (Graciano et al., 2015; Salazar et al., 2015). Thus, the movement of the piston is generated during the power strokes, which allows the crankshaft to be driven. Therefore, the mechanical torque T_{mech} of the diesel engine is generated. The fourth part of the diesel engine, which is the flywheel, where the rotor angular velocity w_r is generated, comprises the engine inertia dynamics, the damping factor (KD), and the loaded alternator. In this sense, the mechanical expression, which relates the rotor angular velocity w_r and the mechanical torque T_{mech} is given by (Graciano et al., 2015; Salazar et al., 2013, 2015):

$$2H\frac{d\Delta w_r}{dt} = T_{mech} - T_{sg} - K_D\Delta w_r \tag{5.1}$$

$$\frac{d\delta}{dt} = w_0\Delta w_r \tag{5.2}$$

$$\Delta w_r = w_r - 1 \tag{5.3}$$

where the time t is in seconds (s), the rotor angle δ is in radian, the rated generator speed w_0 is in rad/s, Δw_r is the speed deviation (pu), w_r is the rotor angular velocity (pu), T_{sg} is the generator torque (pu), and H is the inertia constant.

Synchronous generator

The synchronous generator is modeled using the Park transformation. Hence, the direct and the quadrature voltages can be described as follows (Koutroulis, Kolokotsa, Potirakis, & Kalaitzakis, 2006; Salazar et al., 2013):

$$V_d = E''_d - R_S I_d + X''_q I_q \tag{5.4}$$

$$V_q = E''_q - R_S I_q - X''_d I_d \tag{5.5}$$

where:

R_S: the armature resistance (Ω),
I_d and I_q: the stator direct and the quadrature currents (A), respectively,
$X''_{d,q}$: the sub-transient direct and quadrature reactances, respectively.
$E''_{d,q}$ is expressed by:

$$E''_d = \frac{(X_q - X''_q)}{1 + \tau''_{q0}s} I_q \tag{5.6}$$

$$E''_q = \frac{1}{1 + \tau''_{d0}s} E'_q - \frac{(X'_d - X''_d)}{1 + \tau''_{d0}s} I_d \tag{5.7}$$

where:

$X_{d,q}$ and $X'_{d,q}$: the synchronous and transient reactances,
$\tau''_{d,q0}$: the open circuit sub-transient time constants (s).
E'_q is expressed by:

$$E'_q = \frac{1}{\left(\frac{X_d - X''_d}{X'_d - X''_d}\right) + \tau'_{d0}s} E_{fd} + \frac{\left(\frac{X_d - X'_d}{X'_d - X''_d}\right)}{\left(\frac{X_d - X''_d}{X'_d - X''_d}\right) + \tau'_{d0}s} E''_q \tag{5.8}$$

where:

E_{fd}: the exciter field voltage (V),
τ'_{d0}: the open circuit transient time constant (s).

Excitation system

The excitation system supplies and automatically adjusts the field current of the synchronous generator, by applying control and protective functions required to the

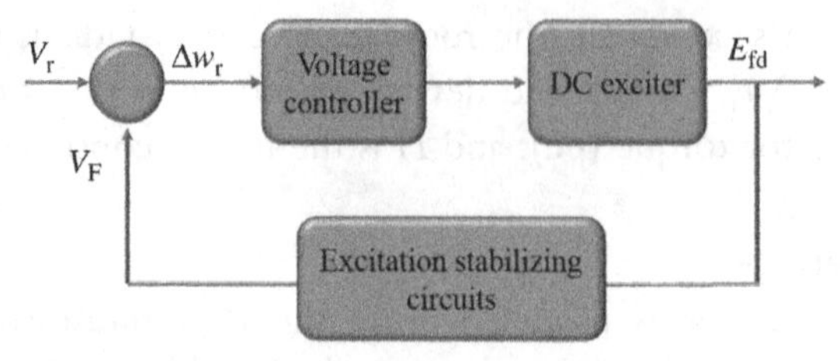

Figure 5.6 Diagram of the excitation system of the synchronous generator.

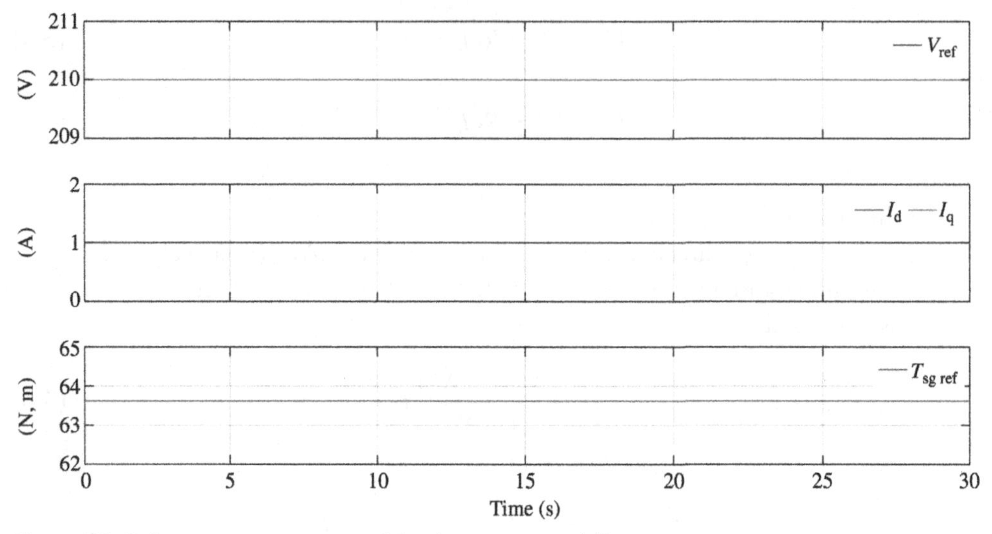

Figure 5.7 Reference parameters used for the genset modeling.

satisfactory performance of the system (Graciano et al., 2015; Salazar et al., 2013, 2015). The functional block diagram of an excitation control system is shown in Fig. 5.6.

5.2.4.2 DG modeling results

The DG model is tested by simulation using Matlab–Simulink. The results are illustrated in Figs. 5.7–5.9, showing the electrical and mechanical variables of the genset.

5.3 ECONOMIC VIABILITY OF PV/BATTERIES AND DIESEL PLANTS

In this section, the total cost including the investments and use costs are compared in three different systems. As it has previously been cited, the first system is composed of a DG, which supplies a water pump. In the second system, the pump is supplied by both a PVP and a DG. Finally, the third system is based on a renewable water pumping plant, in which the pump is supplied by a PVP and a battery bank. The system

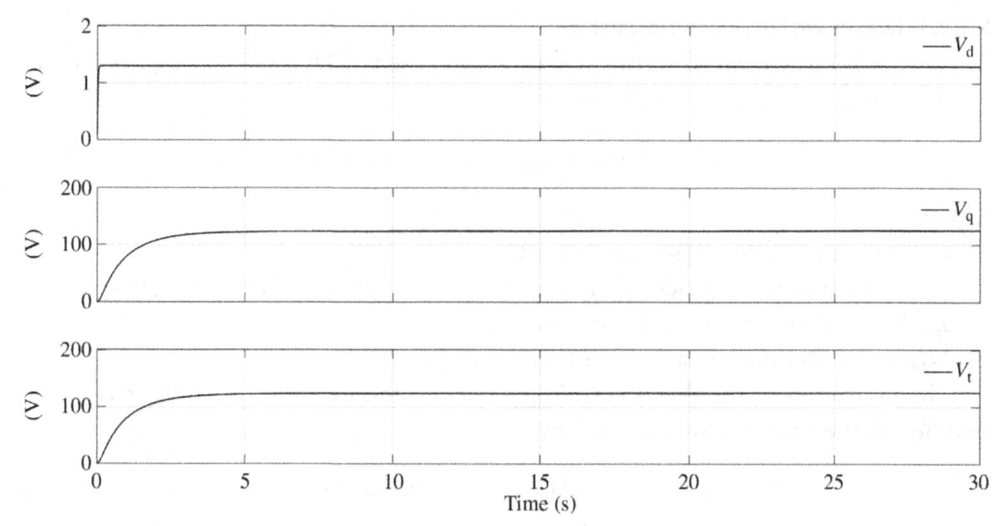

Figure 5.8 DG voltages.

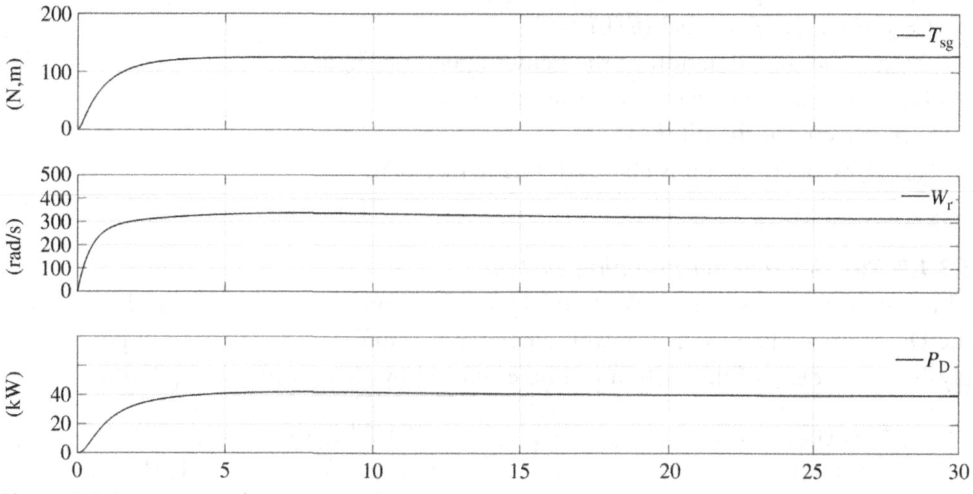

Figure 5.9 Torque, angular speed, and DG power.

which is composed of PV/batteries/DG is not studied here, since it is assumed that the water pumping will be performed only during the day. Hence, the study of these systems costs are detailed now.

5.3.1 Study of the costs of water pumping plants

The costs of a water pumping installations, which are composed of N components, include the investment, the maintenance, and the components replacements costs. They are evaluated for 20 years of operation as follows (Koutroulis et al., 2006).

5.3.1.1 Diesel water pumping system

As it has previously been seen, this system is composed of DG, which supplies a water pump. Hence, its cost $cost_{s1}$ can be evaluated by:

$$cost_{s1} = n_{diesel}(C_{t_{diesel}} + (n_y - 1)\, M_{diesel}) \tag{5.9}$$

where:

n_{diesel}: the number of diesel engines used,
$C_{t_{diesel}}$: the investment price of the diesel engine (€/for n_y years of operating),
n_y: the number of years of the system operation,
M_{diesel}: the DG maintenance cost (€/module per year).

The evaluation of the diesel cost $C_{t_{diesel}}$ includes the fuel and oil costs. It is evaluated, for all the studied systems as follows:

$$C_{t_{diesel}} = C_{diesel} + C_{fuel} * \Delta t_{diesel} * V_{fuel} + n_{oil} * C_{oil} * V_{oil} * n_y \tag{5.10}$$

where:

C_{diesel}: the DG price (€/module for n_y),
C_{fuel}: the cost of the fuel (€/L),
Δt_{diesel}: the time duration of the genset operation (h/day),
V_{fuel}: the volume of the fuel consumption (L/h),
C_{oil}: the cost of the oil (€/L),
n_{oil}: How offen the oil is changed during one year.

5.3.1.2 PV–diesel water pumping system

This system is composed of PVPs and DG, which supply a water pump. In this case, the DG is used when the PV power generated is insufficient for the pump power supply. The cost $cost_{s2}$ of this system can be evaluated by (Koutroulis et al., 2006):

$$\begin{aligned} cost_{s2} = {} & n_{pv}(C_{pv} + n_y M_{pv}) + n_{chop} C_{chop}(y_{chop} + 1) + M_{chop}(n_y - y_{chop} - 1) \\ & + C_{inv}(y_{inv} + 1) + M_{inv}(n_y - y_{inv} - 1) + n_{diesel}(C_{t_{diesel}} + (n_y - 1)M_{diesel}) \end{aligned} \tag{5.11}$$

where:

n_{pv}: the number of PV modules,
C_{pv}: the PV module cost (€/module for n_y),
M_{pv}: the PV module maintenance cost (€/module per year),
n_{chop}: the number of choppers,
C_{chop}: the chopper cost (€/chopper for n_y),
y_{chop}: the number of times the chopper is replaced during n_y years,
M_{chop}: the maintenance cost for one chopper (€/chopper per year),
C_{inv}: the cost of the inverter (€/inverter for n_y),
y_{inv}: the number of the inverter replaced during n_y years,
M_{inv}: the maintenance cost for one inverter (€/inverter per year).

5.3.1.3 PV–battery water pumping system

This system is composed of PVPs and a battery bank, which supply a water pump. In this case, the battery bank is used when the PV power generated is insufficient for the pump power supply. The cost $cost_{s3}$ of this system can be evaluated by (Koutroulis et al., 2006):

$$cost_{s3} = n_{pv}(C_{pv} + n_y M_{pv}) + n_{bat}(C_b + y_b C_b + (n_y - y_b - 1)M_b) + n_{chop} C_{chop}(y_{chop} + 1) + C_{inv}(y_{inv} + 1) + M_{inv}(n_y - y_{inv} - 1) \tag{5.12}$$

where:

n_{bat}: the batteries number,
C_b: the battery cost (€/battery for n_y),
y_{bat}: the number of times the batteries are replaced during n_y years,
M_{bat}: the maintenance cost for one battery (€/battery per year).

5.3.2 Evaluation of water pumping systems costs

The cost parameters for the installations components in the target area (Medjez El Beb, Tunisia) are described in Table 5.1, together with their values selected in the studied water pumping systems.

Table 5.1 Cost parameters for the water pumping installations components [8]

Parameters	Name	Values
n_y (years)	The installation life time	20
C_{pv} (€/module for n_y)	The PV module cost	265.81
M_{pv} (€/module per year)	The PV module maintenance cost	2.66
C_b (€/battery for n_y)	The battery cost	264
y_{bat}	The number of times the batteries are replaced during n_y years	4
M_{bat} (€/battery per year)	The maintenance cost for one battery	2.64
n_{chop}	The number of choppers	1
C_{chop} (€/chopper for n_y)	The chopper cost	200
y_{chop}	The number of times the chopper is replaced during n_y years	0
M_{chop} (€/chopper per year)	The maintenance cost for one chopper	2
C_{inv} (€/inverter for n_y)	The cost of the inverter	1942
y_{inv}	The number of the inverter replaced during n_y years	0
M_{inv} (€/inverter per year)	The maintenance cost for one inverter	19.42
C_{diesel}	The DG cost	4475
M_{diesel} (€/inverter per year)	The maintenance cost for the diesel	44.75

Table 5.2 Parameters of the DG

System	System 1	System 2
Number of hours (h)	13.25	5
Fuel consumption	4.7	L/h
Fuel price	0.48	€
Oil volume	8	L/3 months
1 L oil cost	5.91	€

Table 5.3 Cost evaluation of the three water pumping systems using data of Tunisia

System	System (1): DG/pump	System (2): PV/DG/pump	System (3): PV/batteries/pump
Costs (€)	57,597	65,611	51,263

The evaluation of the DGs cost includes the oil and fuel consumption costs, which are given in Table 5.2.

The costs of the systems above detailed are now evaluated, following the climatic and economic data that correspond to Tunisia, and using the following system components size ($S_{opt} = 101.5\ m^2$ and $n_{bat_{opt}} = 8$ batteries of 210 Ah/12 V), which has been obtained by the sizing algorithm detailed in Section 3.4. Thus, the final costs of the three systems are summarized in Table 5.3.

A significant difference is demonstrated in the costs of the three studied systems using climatic and economic data that correspond to Tunisia. In fact, System (2), which consists of the PV/DG/pump is the most expensive system, since it is based on PVPs and a DG. Indeed, it requires the use of converters (DC–DC and DC–AC) for the solar power control and conversion. Moreover, it is subject for the high price of the DG (considering that the fuel price is constant during the 20 years), which therefore makes its cost the highest among the three studied systems. However, System 1, which is only supplied by the DG, is characterized by a cheaper cost than the cost of System (2). This is explained by the fact that the PVPs and the corresponding converters (which make the total cost high) are not used in System(1). Finally, the cost of System (3), which corresponds to the PV/batteries/pump is the cheapest among the three systems, since the genset is replaced here by a battery bank, which is characterized by a less cost than the one of the DG. Therefore, PV/batteries/pump system is the cheapest solution for the water pumping plant supply in Tunisia. It is necessary to mention that, there is a small difference between System (1) and System (3) costs. However, taking into account that systems which include DG need a continuous maintenance and require the user presence (for fuel supply), which make them a nonpractical solution.

As it has previously been cited in this chapter, this study is performed using geographic and costs data, which correspond to Tunisia. Hence, it is necessary to evaluate whether this study still valid when geographic conditions change. This will be detailed in depth in the next section of this chapter.

5.4 SENSITIVITY ANALYSIS OF THE COST OF WATER PUMPING SYSTEMS TO THE GEOGRAPHIC PARAMETERS

As it has previously been seen, based on climatic, geographic, and economic parameters of Tunisia, the PV/batteries/pump system is the most economic system for water pumping plants in the country. In this section, the economic study, based on the geographic parameters variations, is conducted in depth. Hence, the costs of the three previous systems are detailed in two more countries, Spain (latitude: 40.25°) and Qatar (latitude: 25.16°). The choice of these countries is based on the similarities in the oil price and availability of important and similar solar energy amounts compared of those of Tunisia. For instance, in *July*, the global solar energy evaluated on a tilted PVP in Tunisia, Spain, and Qatar are respectively: 9136.7, 9100, and 8885.2 Wh/m^2.

To perform this study, first, the sizing algorithms presented in Chapter 3, Sizing Optimization of the Photovoltaic Irrigation Plant Components, have been performed for the two countries: Spain and Qatar. Considering that the crops water need is the same, and using climatic data of the three countries, the final PVPs surface and the number of batteries are illustrated in Table 5.4.

It is worth to note here that the obtained sizing results for the three countries are similar, since the global solar energy are similar, from the one hand, and that a same water volume is assumed to be needed in the three countries, from the other one.

Then, the costs of the studied systems in both countries (Spain and Qatar) are evaluated using Eqs. (5.9)−(5.12). To do so, updated fuel prices which correspond to these two countries are used. They are summarized in Table 5.5.

Table 5.4 Sizing of PV−batteries systems using data of Tunisia, Spain, and Qatar

Results	PV surface (m^2)	Batteries number
Country		
Tunisia	101.5	8
Spain	102.5	8
Qatar	104.5	8

Table 5.5 Fuel costs in Tunisia, Spain, and Qatar (global petrol prices website, on August, 12th, 2016)

Country	Tunisia	Spain	Qatar
Costs (€/L)	0.48	0.99	0.34

Table 5.6 Costs evaluation of the three systems using data of Tunisia, Spain, and Qatar

System	DG/pump (1)	PV/DG/pump (2)	PV/batteries/pump (3)
Costs (€)			
Tunisia	57,597	65,611	51,263
Spain	111,130	74,318	50,306
Qatar	42,903	53,553	50,944

The costs results of the three systems, which correspond to DG/pump, PV/DG/pump and PV/batteries/pump systems, evaluated using economic and climatic data of Spain and Qatar are presented in Table 5.6.

The costs obtained using data of Spain show that the DG/pump system (System (1)) represents the most expensive system that can be used for water pumping. Moreover, the inclusion of PV energy in the System (2) makes its price decrease significantly by almost one-third. This is justified by the fact that the Spanish strategy in energy production encourages for renewable energy use and tends to discard fossil energy sources. This is definitely proved by the cost of System (3), which corresponds the cheapest system price, not only compared with the other two systems, but also by comparing it with the cost of the same system type for Tunisia and Qatar. However, the study of the three systems costs for Qatar shows that System 1 is the cheapest system, since in Qatar, the fuel price is very cheap, compared to the fuel costs in Spain and Tunisia (see Table 5.5).

Hence, although there is similarity in the solar energy availability and the oil prices in these three countries, PV/batteries/pump system (System 3) is a good solution only in Tunisia and Spain. However, DG/pump system (System (1)) is the optimum system in Qatar.

5.5 CONCLUSION

This chapter analyzed the cost sensitivity of autonomous water pumping systems based on climatic, geographic, and economic parameters variations. Thus, first, some architectures for water pumping systems are described: they include the DG/pump, the PV/DG/pump, and the PV/batteries/DG/pump, respectively, and a model for the DG is described in depth. Then, the cost of three systems, which are DG/pump, the PV/DG/pump, and the PV/batteries/pump, are evaluated and then compared using climatic, geographic, and economic data of three countries, which are Tunisia, Spain, and Qatar. The obtained results show that PV/batteries/pump system is the cheapest system for both Tunisia and Spain. However, for Qatar, the DG/pump system is the optimum system. Finally, one can conclude that, the fuel price is the basic element, which directs the strategies for renewable or fossil fuel energy sources use for electricity generation. However, the instability of the fuel prices, the decrease in the PVPs price and the enhancements in the PV technologies in the last decades make one believe that fossil fuel sources future for electricity generation is too short.

REFERENCES

Al-Alawi, A., Al-Alawi, S. M., & Islam, S. (2007). Predictive control of an integrated PV-diesel water and power supply system using an artificial neural network. *Renewable Energy, 32*(8), 1426–1439.

Bonanno, F., Consoli, A., Raciti, A., Morgana, B., & Nocera, U. (1999). Transient analysis of integrated diesel-wind-photovoltaic generation systems. *IEEE Transactions on Energy Conversion, 14*(2), 232–238.

Graciano, V., Vargas, J. V., & Ordonez, J. C. (2016). Modeling and simulation of diesel, biodiesel and biogas mixtures driven compression ignition internal combustion engines. *International Journal of Energy Research, 40*(1), 100–111.

Koutroulis, E., Kolokotsa, D., Potirakis, A., & Kalaitzakis, K. (2006). Methodology for optimal sizing of stand-alone photovoltaic/wind-generator systems using genetic algorithms. *Solar Energy, 80*(9), 1072–1088.

Salazar, J., Tadeo, F., De Prada, C., & Palacin, L. (2013). Modeling and simulation of auxiliary energy systems for off-grid renewable energy installations. *Proceeding of IREC congress* (pp. 1–6).

Salazar, J., Tadeo, F., & Prada, C. (2015). Modelling of diesel generator sets that assist off-grid renewable energy micro grids. *Renewable Energy and Sustainable Development, 1*, 72–80.

Wichert, B. (1997). PV-diesel hybrid energy systems for remote area power generation—a review of current practice and future developments. *Renewable Sustainable Energy Reviews, 1*, 209–228.

References

[illegible]

CHAPTER 6

General Conclusion

This book has provided detailed explanations on the modeling, the sizing, the energy management, and the economic study sensitivity of autonomous PV irrigation installations destined for water pumping.

More precisely, first, the system elements models have been detailed and studied in depth. Some of them have been validated experimentally. Then, since the system components have been modeled, it was possible to establish an algorithm for the optimal sizing of the autonomous PV irrigation installation, which is composed of PVPs, a battery bank, an inverter, a water pump, and a reservoir. It has been demonstrated that the proposed algorithm ensures the availability of the required water for irrigation. Moreover, by specifying some criteria related to the use of the battery bank, the sizing algorithm ensures a safe operation for the battery bank. The algorithm and its validation have been performed focusing on a specific case study (tomato irrigation in a region in the north of Tunisia) by using climatic data from the target area during the crops vegetative cycle.

In addition, a fuzzy algorithm for the energy management of the PV installation has been proposed and evaluated. It has been shown that, the algorithm ensures the system autonomy and pumping the water volume needed for the crops irrigation. Moreover, the fuzzy energy algorithm guarantees that the charge in the battery bank is maintained between two fixed values which correspond to a minimum and maximum battery bank *dod*, to protect it from excessive charges and discharges. The algorithm is based on measuring the PV power, the water level in the reservoir, the well water flow, and the battery bank *dod* (estimated from the battery bank current), to deduce the switching of the relays that link the components. The control algorithm reliability has been validated experimentally in a small-scale water pumping plant installed in the laboratory. The results are promising, also making it possible to manage some critical cases, such as the water leak in the reservoir and the case of a discharged battery bank.

Finally, the economic feasibility of using the PV/batteries/pump system has also been studied in depth, by comparing its cost with systems including DG. Then, the study is generalized in two more countries (Spain and Qatar). The obtained results prove that the PV/batteries/pump system is the cheapest system for water pumping

Specifications of Photovoltaic Pumping Systems in Agriculture
DOI: http://dx.doi.org/10.1016/B978-0-12-812039-2.00006-5

systems in Tunisia and Spain, whereas DG/pump system is the optimum solution for water pumping systems in Qatar.

As a general conclusion, in this book, the author presented a complete study of PV energy use in autonomous water pumping systems, in which the energetic and economic criteria and sensitivities have been analyzed in depth.

It has been demonstrated the reliability of PV energy in supplying water pumping systems. Sure, this energy source is getting enhanced following the increasing development of the PV technology. Nevertheless, further work is required to enhance the PV panel yields and to limit phenomena that limit the PV power generation, like soiling and aging. It is thus the authors' hope that the information presented in the book will be useful to the reader.

APPENDIX A: PHOTOVOLTAIC ENERGY: BASIC PRINCIPLES

A.1 INTRODUCTION

The sun is a star which provides nearly 95% of its energy to generate nearly all the energy available on the earth (Mayor & Queloz, 1995). In fact, the earth energy is a direct or indirect conversion form of the sun's lights. For instance, the solar heat ensures the plants evaporation-transpiration by converting the sunlight energy into storable chemical form through the photosynthesis process, which allows the fauna and flora to be maintained in the earth (Romero et al., 2014). Moreover, the sunlight is the source of generating wind and wave energies, by causing temperature difference in the atmosphere. Additionally, sunlight generates the photothermal and photovoltaic energies, which are used to provide electrical and heat energies (Hersch & Zweibe, 1982). Consequently, sun means life.

A.1.1 Composition of light energy

The white sun's light is composed of visible and invisible radiations in interaction, and which are characterized by specific energy and wavelengths (Bisi, 2015). For instance, the red color is the least-energy color in the visible spectrum of the white light. However, the violet color is the highest energy wave of the same spectrum. In the invisible radiations, the infrared region represents the least energy radiations that generate energy in form of heat. While the ultraviolet radiations of the same radiations type (the invisible radiations) has the highest energy, for which, a long human exposure may cause health problems, especially those related with the skin (James, Berger, & Elston, 2015). Indeed, the ultraviolet region is characterized by a short distance between its wave peaks (Eskizeybek, Sari, & Gülce, 2012). Thus, it has a higher frequency and generates a higher energy than a shorter wavelength wave (namely the infrared radiations). Consequently, the wave energy is directly proportional to the wave frequency (Sun, Timurdogan, & Yaacobi, 2013).

A.1.2 Sunlight on the earth

The solar radiation that reaches the earth is approximated to 1.37 kilowatts per square meter (kW/m^2) (Kalogirou, 2013). This radiation amount is considered low, compared with solar radiation generated by the sun. This is justified by the fact that not all the sunlight incident on the earth's atmosphere can reach the earth's surface and that a big amount of the sunlight is lost before reaching the earth. In fact, several phenomena act together to attenuate the sunlight. First, the atmosphere itself decreases the sunlight. Indeed, during this process, some rays are totally absorbed, such as the X-rays, filtered like the ultraviolet radiation, or reflected back into the space. Additionally, some waves of the sunlight are randomly scattered by the atmosphere, which therefore gives the blue color to the sky (Fabelinski, 2012). Moreover, the air quantity, which is known also as the air mass (AM), acts to attenuate the sunlight (Zhang, Wang, & Xu, 2013). For instance, AM0 is attributed to zones where there is no air, like zone before the atmosphere. AM1 is associated to tropical regions, which are exposed to the sun radiations for long hours of the day and during long periods of the year. Consequently, the more the AM is high, the less the site is exposed to the sunlight.

After passing the atmosphere, the sunlight strikes the earth depending on the site latitude. In fact, areas located between 30° north and 30° south latitude, namely tropical zones, are more likely to receive more sunlight than other regions through the year. This is justified by the fact that these areas have the longest day hours, from the one hand, and that the angles between the sun radiations and these sites is high, from the other, which therefore favorites the increase of the time duration and the earth surfaces stroked by sunlight (Mettanant & Chaiwiwatworakul, 2014).

In addition to the dependency on the latitude, thanks to the earth rotation around the sun and its declination angle with respect to the sun, this creates seasonal variations of the sun radiation. Consequently, this results in the temperature variation and therefore having seasons over the year (Mettanant & Chaiwiwatworakul, 2014). For instance, areas located at latitudes higher than 30° north and more than 30° south are more likely to have temperature variability over the year, since the sunlight reaching these areas varies during the year (Mettanant & Chaiwiwatworakul, 2014). Whereas, locations whose latitudes are between 30° north and 30° south, they are have tropical climate, which is generally characterized by rain and no rain seasons, and also high average temperatures (Mettanant & Chaiwiwatworakul, 2014). In areas that have four seasons (winter, spring, summer, and autumn), during winter, the sun provides less than 20% of the summer sun's energy, since the sun is lower in the sky, from the one hand, and the daylight hours are shorter, from the other. Hence, sunlight that strikes these zones is variable during the year (Laundal, Finlay, & Olsen, 2016).

A.1.3 Photovoltaic energy history

In 1839, a 19-year-old French physicist named Edmund Becquerel discovered the ability of some materials, named semiconductors, to generate charge carriers from light, which was after that named Photovoltaic effect (Ghosh, Biswas, & Balaji, 2015).

The French physicist succeeded in generating a voltage using the semiconductor material by illuminating a metal electrode in a weak electrolyte solution (Ghosh et al., 2015). Forty years later, the scientists Adams and Day used Selenium to generated electricity and reached 1–2% cell efficiency (Green, 2015). Then, in 1905, Albert Einstein explained the photoelectric effect in sevral publications (Yu et al., 2015). Later, by the years 1940s and 1950s, the Polish scientist Czochralski made the first technique to generate single crystal silicon used for photovoltaics, which continues to dominate the photovoltaic industrial techniques up to today (Lin et al., 2016). Then, researches have been focusing in the enhancement of silicon efficiency, such as works developed at Bell Telephone Laboratories, which designed a silicon photovoltaic cell with a 4% efficiency, that have been enhanced to 6% and then to 11%, heralding an entirely new era of PV cells (Nayak, Garcia-Belmonte, Kahn, Bisquert, & Cahen, 2012).

The development and the commercialization of PV cells succeeded, in 1958, when PV cells have been used for space applications, due to the lack of energy sources in the space (Myers et al., 2016). Additionally, the design of the transistor contributed significantly to solar cell technology development, since transistors and PV cells are made from similar materials. Then, by the 1960s, new materials have been used to design PV cells, namely gallium arsenide (GaAs), which is characterized by its ability to work under higher temperatures than silicon.

Nowadays, researches are more focused to study the PV end use causes and phenomenon that affect the PV cells operation, like the shading and soiling effects. This is because PV energy is a promote energy source, especially with the instability of fossil fuel price. Hence, PV energy is considered cost effective, especially in areas that are remote from utility grids, where the power supply from conventional sources is impractical, due to the geographic difficulties to extend the electric grid there. For grid-connected systems, the actual cost of photovoltaic electricity is considered high, since power is provided during periods of peak demand, thereby, it decreases the use of storage systems, to cover the peak demand. In addition, PV plants are generally installed near to the places of consumption, to decrease the installation costs, reduce transmission and distribution losses, and increase the system reliability. Consequently, solar energy is abundant, inexhaustible, and clean; yet, PV energy requires special techniques to gather enough of solar energy effectively.

A.2 PHOTOVOLTAIC EFFECT PRINCIPLE

A.2.1 Description of the silicon atom

Silicon atom has 14 electrons arranged in such a way that the outer 4 electrons, which are called valence electrons, can be given to, accepted from, or shared with another atom, which can be silicon or not (Pla et al., 2012; Wolfowicz et al., 2013). These valence atoms are generally used to link silicon atoms together, and hence forming tetrahedral arrangements named crystals, where the silicon atoms are placed centered in the cubic faces (Yun et al., 2013).

A.2.2 Effect of light on silicon atoms

The light that reaches a silicon atom is classified into two types following the energy level to low and high energy lights (Wenham, 2012). Indeed, when a low energy light is detected by a silicon atom, this generates vibration in the silicon electrons in their bound positions. Thus, they gain more energy but it is not sufficient to make the electrons move from their bound position. Consequently, the electrons return to their original lower energy levels, giving off as heat the energy they had previously acquired. However, light characterized by higher energy level can alter the electrical properties of the silicon atom, in such a way that the electron can be torn from its place in the crystal, leaving behind a silicon bond that is missing an electron, and which is named a hole, and moving to the crystal's conduction band.

In this sense, the generation of charge carriers, which are electrons and holes by light is the central process in the overall PV effect. However, since electrons will be recombined with halls and only heat will be obtained, it is necessary to find a mechanism which ensures avoiding the recombination phenomena happening, from the one hand, and allows exploiting the electrons and holes to produce an electrical energy. This can be performed using a potential barrier, which is explained now (Tamura & Burghardt, 2013).

A.2.3 Potential barrier principle

a. *Barrier Operation Principle*

To avoid the charge carriers recombination, potential barriers are conceived in the PV cells, so as to separate the electrons and holes, which are generated by the sunlight, by arranging them properly in such a way to have each charge carrier type in one different side (Sharma, 2013). Consequently, the voltage at the silicon cell ends can be generated, and therefore, it can be later used to generate an electric current in an external specific circuit.

b. *Barrier Forming Principle*

Forming the barrier requires the presence of two types of polarities in the cell: negative and positive. To obtain them, silicon atoms are generally doped by introducing impurities, at very high temperatures (up to 700°C), into the silicon crystal (Edri, Kirmayer, Kulbak, Hodes, & Cahen, 2014). Generally, the atom of bohr (B), which has three valence electrons and the atom of phosphorus (P), which has five valence electrons are the most famous atoms used to dope the silicon crystal

(Lee, Lin, & Lin, 2016). Since the phosphorous (P) has five valence electrons, which means an extra electron compared with the silicon valence electrons number, thus it is used for N-type doping, whereas, the bohr (B) is used for P-type doping, since it has a leak of an electron, compared with the silicon valence electrons number, which has four valence electrons. The doping technique can significantly increase the silicon atoms conductivity by a factor of 10^6 (Edri et al., 2014).

c. *Principle of P–N Junction Operating*

The doping of Silicon atoms with the bohr and the phosphor atoms allows having two types of doped silicon, P-type and N-type, respectively. Thus, using these two doped Silicon types, the zone that separates the P-type and the N-type is called P-N junction, in which, a difference in the potential is created. In fact, in the N-type semi-conductor, there are an excess of electrons in the valence bond, whereas, in the P-type semi-conductor, there is an excess of holes, and thus a leak of electrons in the valence bond. Therefore, when the N-type-doped silicon is in contact with a P-type-doped Silicon; this allows electrons in excess, including valence electrons to move to the P-material, and consequently, the holes in excess to move to N-type material. This creates a charge carrier imbalance (Lee et al., 2016). However, the charges carriers that jumped the P-N junction recombine and are filled with each other. This phenomenon is continuous, and creates a voltage at the P–N junction (Christesen et al., 2012).

Hence, when a sun light characterized by sufficient energy is detected by a PV cell, an electron–hole pair is generated in the P–N junction (Haberlin & Eppel, 2012). Thus, taking advantage from the barrier voltage at the P–N junction, the electron is accelerated to the N-type area. This electron cannot move to the P-type area, thanks to the voltage in the P–N junction. Moreover, there is low probability that it recombines with a hole in the P–N junction, due to the low number of holes existing there. Similarly, its hole partner, is accelerated by the voltage existing in the P–N junction toward the P-area material. Since there are few electrons at the P-type area, the recombination phenomena is limited. Consequently, the P–N junction allows the charge carriers to be immigrated each one to an area type (Haberlin & Eppel, 2012).

The connection of the two silicon material types (P and N) to an external circuit, allows the electric current to be circulated, from the one hand, and the load to be supplied, from the other (Sarver, Al-Qaraghuli, & Kazmerski, 2013). Therefore, when a photon strikes a PV cell with the sufficient energy, its energy is ceded to electron–hole pairs. Consequently, the more the solar radiance is detected by the PV cell, the more electric current is generated. In this sense comes the importance of choosing the optimum PV panels' location and tilted angle when installing it (Sarver et al., 2013).

A.3 PHYSICAL ASPECTS OF SOLAR CELL EFFICIENCY

Generally, the efficiency of the PV cells is approximated to be between 10% and 15%. This low amount is owing to several phenomena, which decreases the amount of converted sunlight to electric current. Among them, the sunlight reflection causes the loss of approximately 36% of the solar radiations that are detected by the PV cell. Another reason for low efficiency of PV cells is that sunlight with low energy cannot create electron–hole pair, or in case that the charge carriers are generated, they finish by recombining rapidly and their energy is lost in heat form (Hersch & Zweibe, 1982). Additionally, PV cell losses are also associated to light that is too energetic for the generation of an electron hole pair. Joined, these two phenomenon are responsible of losing around 55% of the energy from the original sunlight (Hersch & Zweibe, 1982). The recombination of the charges carriers is considered a main reason for the PV cell low efficiency, since the electrons and the holes lose their energy in form of heat energy. In addition, anomalies in the PV cell fabrication resulted from bad contacts between the PV cells and the electrical conductors that collect electrical current lead to the high resistance of the material, which decrease the capacity of PV cells in generating electrical energy (Hersch & Zweibe, 1982). Moreover, others external factors affect the PV cell operation, namely the shading and the soiling effects. Indeed, the shading effect phenomenon may be resulted by shades created by trees, buildings, clouds, or the PV modules themselves between each other. Shading causes a nonuniformity in the distribution of the solar radiation in the PV modules and/or the PV substrings, and therefore, the shaded PV module and/or PV substring limits the PV power generated by the whole PV panels (Sullivan, Awerbuch, & Latham, 2013). The soiling decreases the solar radiation that can be detected by the PV cells, depending on the installation' site characteristics (rural, urban, Sahara), and the climatic characteristics (wind humidity, rainfall) (Hersch & Zweibe, 1982). Furthermore, high ambient temperatures also cause the degradation and low efficiency for the PV cells, since it affects directly the charge carrier energy. However, these losses depend on the PV cell material. For instance, for monocrystalline silicon, the losses in the PV cells are estimated to be 0.4%, while they are 0.11 % for amorphous silicon (Sullivan et al., 2013).

A.4 SOLAR CELL MAKING TECHNIQUE

As it is well known, silicon is obtained after performing a specific treatment to quartzite, which is the most abundant material in the earth, used to make several components including electronic components like transistors and solar cells (Hersch & Zweibe, 1982). However, since this material is highly impure, it is necessary to exclude impurities from Quartzite, to be able to be used later for PV cell fabrication.

In fact, one of the techniques used to exclude the impurities consists first in heating the quartzite in presence of the carbone, in such a way to break the SiO_2 into elemental silicon and carbondioxide. Then, a specific gas is applied for these elements, which is characterized by its specific reaction only with the impurities. This gas allows the impurities to be disappeared in form of gas. Nevertheless, some of the impurities still exist; hence, chemical material is used to make the silicon converted into liquid called trichlorosilane, which is then distilled, to separate silicon from the impurities (Hersch & Zweibe, 1982). Finally, the compound $SiHCl_3$ is broken down and the pure silicon is isolated by a chemical vapor deposition (Hersch & Zweibe, 1982). This technique allows having a polycrystalline silicon wafers. Therefore, to obtain a monocrystalline silicon, its atoms must be arranged in a perfect lattice, when the Si atoms solidification is performed during enough time during the cooling process. Consequently, once the polycrystalline or the monocrystalline silicon is fabricated, it is cut into wafers using the desired thick and diameters (Hersch & Zweibe, 1982).

A.5 PV ARRAYS

PV cells produce low power. Thus, to have electric current and voltage sufficient to supply plants, they must be connected between each other, to form PV modules, which are also connected in series and/or in parallel, depending on the load characteristics, to form PV panels or arrays (Sullivan et al., 2013). However, these connections (series/parallel) must not affect the whole system resistance over climatic parameters. In fact, depending on the installation site, the PV panels must be able to face mechanical forces, namely winds and the site soiling effects. These criteria must be taken into account, especially for systems that are equipped with sun trackers, which are used to maximize the power acquired from the PV panels (Sullivan et al., 2013). In fact, these trackers can have one direction (N to S or E to W) or two directions. The first sun tracker type allows tracking the sun elevation in the sky, which varies considerably each month. However, the second sun tracker type allows the sun to be tracked hourly during the daylight, following the azimuth. Additionally, other technique are used to enhance the PV panels power generation, namely the use of mirrors, which ensure reflecting to the PV panels the sunrays. Additionally, as it has been previously seen, the high temperatures decrease the PV panels efficiencies (Hersch & Zweibe, 1982). Generally, natural air serves for cooling PV panels, if they are not installed in a manner that impedes the air circulation. In addition, water-cooling tubes can be installed at the back of the PV panels and used to cool the PV panels, from the one hand, and to obtain hot water, from the other one (Sullivan et al., 2013).

REFERENCES

Bisi, O. (2015). *Visible and Invisible.* Springer.

Christesen, J. D., Zhang, X., Pinion, C. W., Celano, T. A., Flynn, C. J., & Cahoon, J. F. (2012). Design principles for photovoltaic devices based on Si nanowires with axial or radial p–n junctions. *Nano letters, 12*(11), 6024–6029.

Edri, E., Kirmayer, S., Kulbak, M., Hodes, G., & Cahen, D. (2014). Chloride inclusion and hole transport material doping to improve methyl ammonium lead bromide perovskite-based high open-circuit voltage solar cells. *The Journal of Physical Chemistry Letters, 5*(3), 429–433.

Eskizeybek, V., Sari, F., & Gülce, H. (2012). Preparation of the new polyaniline/ZnO nanocomposite and its photocatalytic activity for degradation of methylene blue and malachite green dyes under UV and natural sun lights irradiations. *Applied Catalysis B: Environmental, 119*, 197–206.

Fabelinski, I. L. (2012). *Molecular scattering of light.* Springer Science & Business Media.

Ghosh, D., Biswas, K., & Balaji, S. (2015). A revisit on solar cell: generation of electricity by harvesting sunlight. *Science and Culture, 81*(11–12), 337–347.

Green, M. A. (2015). *Forty years of photovoltaic research at UNSW. Journal and Proceedings of the Royal Society of New South Wales* (p. 2). Royal Society of New South Wales.

Haberlin, H., & Eppel, H. (2012). *Photovoltaics.* Wiley.

Hersch, P., & Zweibe, lK. (1982). *Basic photovoltaic principles and methods.* Golden, CO (USA): Solar Energy Research Inst.

James, W. D., Berger, T., & Elston, D. (2015). *Andrews' diseases of the skin: clinical dermatology.* Elsevier Health Sciences.

Kalogirou, S. A. (2013). *Solar energy engineering: processes and systems.* Academic Press.

Laundal, K. M., Finlay, C. C., & Olsen, N. (2016). Sunlight effects on the 3D polar current system determined from low Earth orbit measurements. *Earth, Planets and Space, 68*(1), 142.

Lee, C.-I., Lin, Y.-T., & Lin, W.-C. (2016). Small-signal modeling for pn junctions in the breakdown region at different temperatures using artificial neural networks. *Journal of Electromagnetic Waves and Applications*, 1–8.

Lin, Y., Xu, Z., Yu, D., Lu, L., Yin, M., Tavakoli, M. M., ... Li, D. (2016). Dual-layer nanostructured flexible thin-film amorphous silicon solar cells with enhanced light harvesting and photoelectric conversion efficiency. *ACS applied materials & interfaces, 8*(17), 10929–10936.

Mayor, M. & Queloz, D. (1995). A Jupiter-mass companion to a solar-type star.

Mettanant, V., & Chaiwiwatworakul, P. (2014). Automated vertical blinds for daylighting in tropical region. *Energy Procedia, 52*, 278–286.

Myers, M. G., Piszczor, M. F., Krasowski, M. J., Prokop, N. F., Wolford S. D. & McNatt, J. S. (2016). "Further Analyses of the NASA Glenn Research Center Solar Cell and Photovoltaic Materials Experiment onboard the International Space Station". In 14th International Energy Conversion Engineering Conference 4930.

Nayak, P. K., Garcia-Belmonte, G., Kahn, A., Bisquert, J., & Cahen, D. (2012). Photovoltaic efficiency limits and material disorder. *Energy & Environmental Science, 5*(3), 6022–6039.

Pla, J. J., Tan, K. Y., Dehollain, J. P., Lim, W. H., Morton, J. J., Jamieson, D. N., ... Morello, A. (2012). A single-atom electron spin qubit in silicon. *Nature, 489*(7417), 541–545.

Romero, E., Augulis, R., Novoderezhkin, I. V., Ferretti, M., Thieme, J., Zigmantas, D., & Van Grondell, R. (2014). Quantum coherence in photosynthesis for efficient solar-energy conversion. *Nature physics, 10*(9), 676–682.

Sarver, T., Al-Qaraghuli, A., & Kazmerski, L. L. (2013). A comprehensive review of the impact of dust on the use of solar energy: History, investigations, results, literature, and mitigation approaches. *Renewable and Sustainable Energy Reviews, 22*, 698–733.

Sharma, B. L. (Ed.), (2013). *Metal-semiconductor Schottky barrier junctions and their applications* Springer Science & Business Media.

Sullivan, C. R., Awerbuch, J. J., & Latham, A. M. (2013). Decrease in photovoltaic power output from ripple: Simple general calculation and the effect of partial shading. *IEEE Transactions on Power Electronics, 28*(2), 740–747.

Sun, J., Timurdogan, E., & Yaacobi, A. (2013). Large-scale nano photonic phased array. *Nature, 493* (7431), 195–199.

Tamura, H., & Burghardt, I. (2013). Potential barrier and excess energy for electron–hole separation from the charge-transfer exciton at donor–acceptor heterojunctions of organic solar cells. *The Journal of Physical Chemistry C, 117*(29), 15020–15025.

Wenham, S. R. (2012). *Applied photovoltaics*. Routledge.

Wolfowicz, G., Tyryshkin, A. M., George, R. E., Riemann, H., Abrosimov, N. V., Becker, P., ... Morton, J. J. L. (2013). Atomic clock transitions in silicon-based spin qubits. *Nature nanotechnology, 8* (8), 561–564.

Yu, S., Frisch, J., Opitz, A., Cohen, E., Bendikov, M., Koch, N., & Salzmann, I. (2015). Effect of molecular electrical doping on polyfuran based photovoltaic cells. *Applied Physics Letters, 106*(20), 203–301.

Yun, G., Kim, K.-M., Roh, Y., Min, Y., Lee, J.-K., & Kim, Y. H. (2013). *Comparison of slowness curves of Lamb wave with elastic moduli and crystal structure in silicon wafers. 2013 IEEE International Ultrasonics Symposium (IUS) 1598-1601.* IEEE.

Zhang, M., Wang, Y., & Xu, M. (2013). Design of high-efficiency organic dyes for titania solar cells based on the chromophoric core of cyclopentadithiophene-benzothiadiazole. *Energy & Environmental Science, 6*(10), 2944–2949.

APPENDIX B: CENTRIFUGAL WATER PUMPS: BASIC PRINCIPLES

B.1 PRINCIPLE OF OPERATION

The centrifugal pump is a kind of pump that converts energy of a prime mover of an electric machine into velocity in the pump impeller, or kinetic energy that allows energy to be ceded to the water, via the generation of a specific pressure in the pump volute. In fact, its operation consists in the fact that the water enters to the pump impeller center. Then, it is accelerated centrifugally towards the pump volute, thanks to the presence of vanes inside (Sahdev, 2004). Therefore, the pumped water head H depends on the water velocity at the impeller periphery. It is expressed by (Sahdev, 2004):

$$H = \frac{v^2}{2g} \tag{B.1}$$

where

H: total head (ft),

g: gravity acceleration (ft/s^2),

v: water velocity at the impeller (ft/s), and it is expressed by (Sahdev, 2004):

$$v = \frac{N\,D}{229} \tag{B.2}$$

where

N: the impeller revolution (RPM),

D: the impeller diameter (inches).

Indeed, the water head should be able to browse the total manometric high (HMT) needed to pump the water from a reservoir to a tank. The HMT is expressed by:

$$HMT = H_g + \Delta J \tag{B.3}$$

where

H_g: the geometric head between the water surface and the water output (m). It is expressed by:

$$H_g = H_a + H_p \tag{B.4}$$

where

H_a: the geometric suction head (m),

H_p: the geometric water discharge high (m),

ΔJ: the water losses due to the water frictions with the conductors (m[3]). It can be expressed by (Moreno, Planells, Córcoles, Tarjuelo, & Carrión, 2009):

$$\Delta J = \frac{10.674\ L\ Q^{1.852}}{C^{1.852}\ D_{\text{int}}^{4.871}} \quad \text{(B.5)}$$

where

L: the length of the water conductor (m),

Q: the water flux (m^3/s),

C: Hazen-Williams coefficient (150 for the PVC),

D_{int}: the conduct diameter (m).

The pumped water flow Q depends generally on the liquid density, the pump size, the rotational component (impeller) velocity, the operational pressure, and the ambient temperature. It is expressed by (Sahdev, 2004):

$$Q = 449\ v\ A \quad \text{(B.6)}$$

where

A: the water conductor area (ft^2).

The pump input power BHP depends on the total head H and the water weight, which is pumped during Δt. It is expressed by (Sahdev, 2004):

$$BHP = \frac{Q\ H\ G_s}{c_t\ \eta_{\text{pump}}} \quad \text{(B.7)}$$

where

G_s: the specific water gravity,

η_{pump}: the pump efficiency (%).

c_t: a coefficient that describes the foot-pounds for one horsepower (33,000) divided by the weight of one gallon of water (8.33 pounds).

A characteristic for centrifugal pumps is that they satisfy the similitude laws. This means that, knowing the rotation speed N_n of the impeller at the instant n, it is possible then to deduce its speed N_{n+1} at $n+1$. Thus, using the water flow Q_n, the total manometric high HMT_n and the absorbed power BHP_n and the variation of the rotation number, the water flow Q_{n+1} at $n+1$ is expressed by (Moreno et al., 2009):

$$Q_{n+1} = Q_n \frac{N_{n+1}}{N_n} \quad \text{(B.8)}$$

Similarly, the water head H_{n+1} at the instant $n+1$ is proportional to the square of the rotation number quotient, as it is described as follows (Moreno et al., 2009):

$$H_{n+1} = H_n \left(\frac{N_{n+1}}{N_n} \right)^2 \tag{B.9}$$

Therefore, the cube of the rotation number quotient gives the absorbed power $P_{\text{pump}\ n+1}$ at the instant $n+1$. It is expressed as follows (Moreno et al., 2009):

$$P_{\text{pump}\ n+1} = P_{\text{pump}\ n} \left(\frac{N_{n+1}}{N_n} \right)^3 \tag{B.10}$$

REFERENCES

Moreno, M. A., Planells, P., Córcoles, J. I., Tarjuelo, J. M., & Carrión, P. A. (2009). Development of a new methodology to obtain the characteristic pump curves that minimize the total cost at pumping stations. *Biosystems Engineering*, *102*(1), 95–105.

Sahdev, M. (2004). "Centrifugal Pumps: Basic Concepts of Operation, Maintenance, and Troubleshooting. Part I", Presented at The Chemical Engineers' Resource Page, www.cheresources.com.

APPENDIX C: THE INDUCTION MACHINE: MODELING AND CONTROL

C.1 SPACE VECTOR NOTION

The space vector $\overline{x}$ is defined by (Fig. C.1):

$$\overline{x} = \frac{2}{3}\left(x_1 + ax_2 + a^2x_3\right)$$

where

$$a = e^{j\frac{2\pi}{3}}$$

and

$$\begin{cases} x_1 = X_1 \cos\left(wt + \varphi_1\right) \\ x_2 = X_2 \cos\left(wt + \varphi_2\right) \\ x_3 = X_3 \cos\left(wt + \varphi_3\right) \end{cases} \tag{C.1}$$

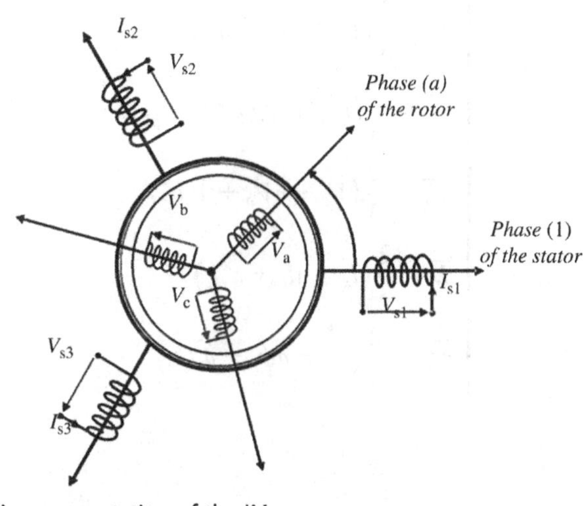

Figure C.1 Schematic representation of the IM.

C.2 REFERENCE CHANGE

In the reference $(\vec{O}x, \vec{O}y)$, the variable $\overline{X}$ can be expressed by $\overline{X_n}$, where $\overline{X_n} = Xe^{-j\beta}$ and $\frac{d\beta}{dt} = w_s$.

This change allows the stator and rotor variable to be constant in the permanent state using a stator reference in the stator field (Fig. C.2).

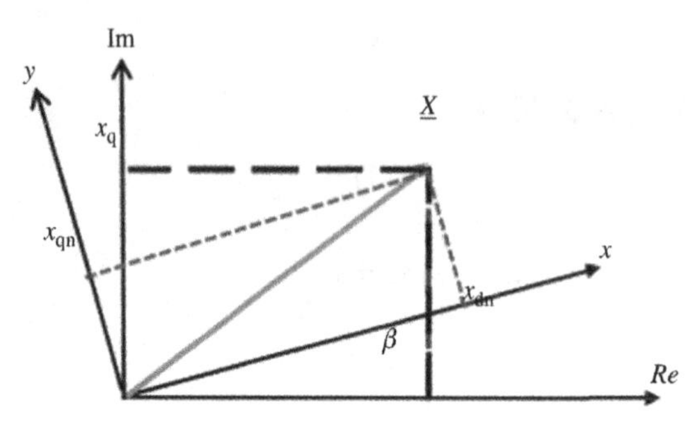

Figure C.2 Representation of the used reference frame.

C.3 EQUATIONS OF THE IM USING THE SPACE VECTORS

Using the vector transformation for the voltage currents and fluxes vectors in the stator and the rotor, we obtain (Sallem, 2009):

$$\begin{cases} \overline{V_s} = \frac{2}{3}\left(V_{s1} + aV_{s2} + a^2V_{s3}\right) \\ \overline{I_s} = \frac{2}{3}\left(I_{s1} + aI_{s2} + a^2I_{s3}\right) \\ \overline{\psi_s} = \frac{2}{3}\left(\psi_{s1} + a\psi_{s2} + a^2\psi_{s3}\right) \\ \overline{V_r} = \frac{2}{3}\left(V_a + aV_b + a^2V_c\right) \\ \overline{I_r} = \frac{2}{3}\left(I_a + aI_b + a^2I_c\right) \\ \overline{\psi_r} = \frac{2}{3}\left(\psi_a + a\psi_b + a^2\psi_c\right) \end{cases} \quad \text{(C.2)}$$

The electric and magnetic equations that describe the IM operation can be expressed by (Sallem, 2009):

$$\begin{cases} \overline{V_s} = R_{ss}\overline{I_s} + \dfrac{d}{dt}\overline{\psi_s} \\ \overline{V_r} = R_r\overline{I_r} + \dfrac{d}{dt}\overline{\psi_r} \end{cases} \tag{C.3}$$

$$\begin{cases} \overline{\psi_s} = L_s\overline{I_s} + me^{j\theta}\overline{I_r} \\ \overline{\psi_r} = L_r\overline{I_r} + me^{j\theta}\overline{I_s} \end{cases} \tag{C.4}$$

The electromagnetic torque C_{em} is expressed by (Sallem, 2009):

$$C_{em} = \frac{3}{2}m\mathrm{Im}\left(\overline{I_s}\left(\overline{I_r}\, e^{j\theta}\right)^*\right) \tag{C.5}$$

The mechanic equation is expressed by (Sallem, 2009):

$$\frac{d}{dt}w_m = \frac{d^2}{dt^2}\theta = \frac{1}{J}p(C_{em} - C_r) \tag{C.6}$$

C.4 STATE EQUATIONS

Using Eqs (C.4–C.6) and using the reference field related to the stator, the rotor variables are (Kamoun, 2011): $\overline{I'_r} = \overline{I_r}\, e^{j\theta}$, $\overline{\psi'_r} = \overline{\psi_r}\, e^{j\theta}$ and $\overline{V'_r} = \overline{V}_r\, e^{j\theta}$.

The electric equations for the stator and rotor circuits can be expressed by (C.7) (Kamoun, 2011):

$$\begin{cases} \overline{V_s} = R_{ss}\overline{I_s} + \dfrac{d}{dt}\overline{\psi_s} \\ \overline{V'_r} = R_r\overline{I'_r} + \dfrac{d}{dt}\overline{\psi'_r} - jw_r\overline{\psi'_r} \end{cases} \tag{C.7}$$

The magnetic equations following the stator reference is expressed by (Kamoun, 2011):

$$\begin{cases} \overline{\psi_s} = l_s\overline{I_s} + M\overline{I'_r} \\ \overline{\psi'_r} = l_r\overline{I'_r} + M\overline{I_s} \end{cases} \tag{C.8}$$

The mechanic equation becomes (Kamoun, 2011):

$$J\frac{d^2\theta}{dt} = J\frac{dw_m}{dt} = C_{em} - C_r = \frac{3}{2}m\mathrm{Im}\left(\overline{I_s}\left(\overline{I'_r}\right)^*\right) - C_m \tag{C.9}$$

To obtain the state equations of the IM, the fluxes have been chosen as state variables. Then, the changing the complex differential system obtained to differential system with real coefficients, an expression that relates the fluxes to currents can be obtained and it is given by Eq. (A.10) (Kamoun, 2011):

$$\overline{\psi} = L\overline{I} \tag{C.10}$$

where

$$\overline{\psi} = \begin{bmatrix} \psi_{sd} + j\psi_{sq} \\ \psi'_{rd} + j\psi'_{rq} \end{bmatrix}, \ \overline{I} = \begin{bmatrix} I_{sd} + jI_{sd} \\ I'_{rd} + jI'_{rq} \end{bmatrix}, \ \overline{V} = \begin{bmatrix} V_{sd} + jV_{sq} \\ V'_{rd} + jV'_{rq} \end{bmatrix}, \ L = \begin{bmatrix} l_s & m \\ m & l_r \end{bmatrix}$$

C.5 IM DIRECT STARTING

In this paragraph, the stator currents, the speed, and the electromagnetic torque that correspond to a direct start-up for the IM are presented (Fig. C.3).

A direct start-up for the IM shows that the stator current is high and may reach four times the nominal value. Hence, the use of a control method is needed.

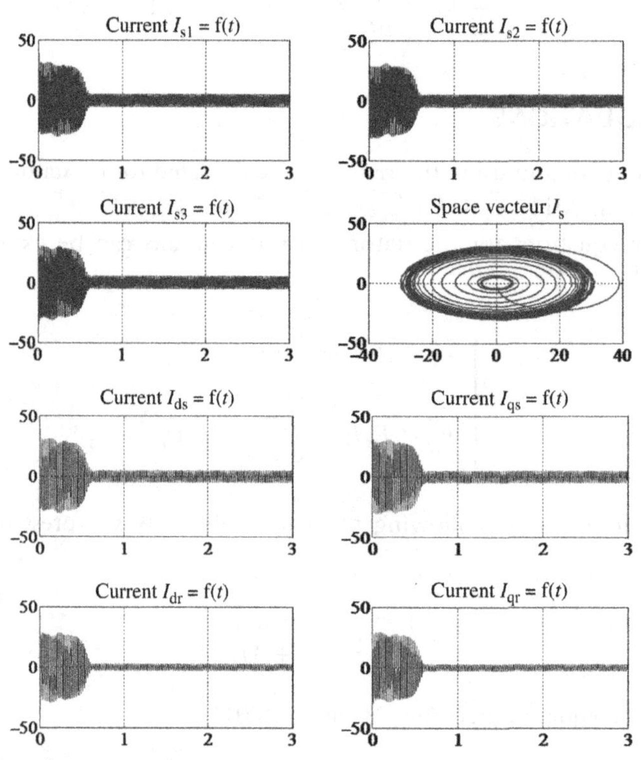

Figure C.3 IM Direct Start-up results.

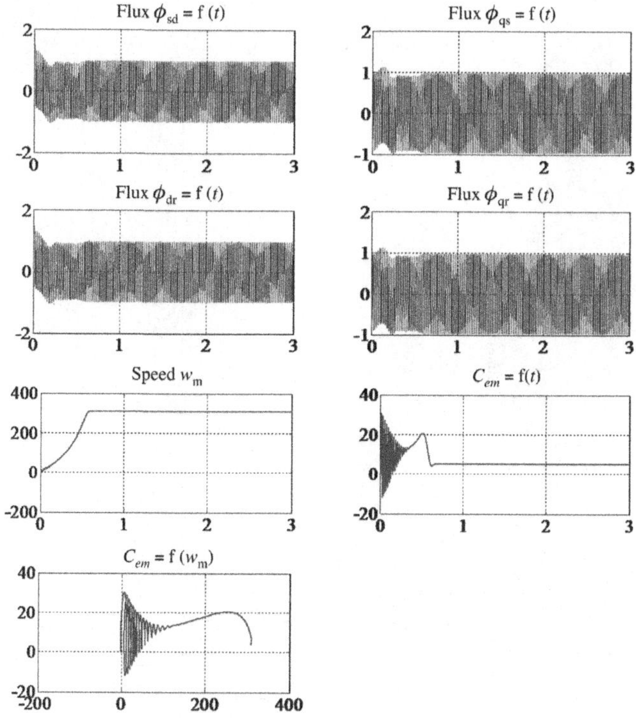

Figure C.3 Continued.

Table C.1 IM parameters (Sallem, 2009)

Parameters	Values
R_{ss}	5.72 Ω
R_{rr}	4.2 Ω
L_s	0.462 H
L_r	0.462 H
M	0.44 H
p	2
J	0.0049 kg.m^2

C.6 IM CONTROL USING THE RFOC METHOD

The IM model is tested using the vector control with the RFOC method (Yahyaoui, Sallem, Chaabene, & Tadeo, 2012). Hence, in this simulation, it is supposed that the pump is supplied only by the photovoltaic panel. The IM parameters are given by Table C.1. The simulation results are given in Fig. C.4 (Yahyaoui et al., 2012).

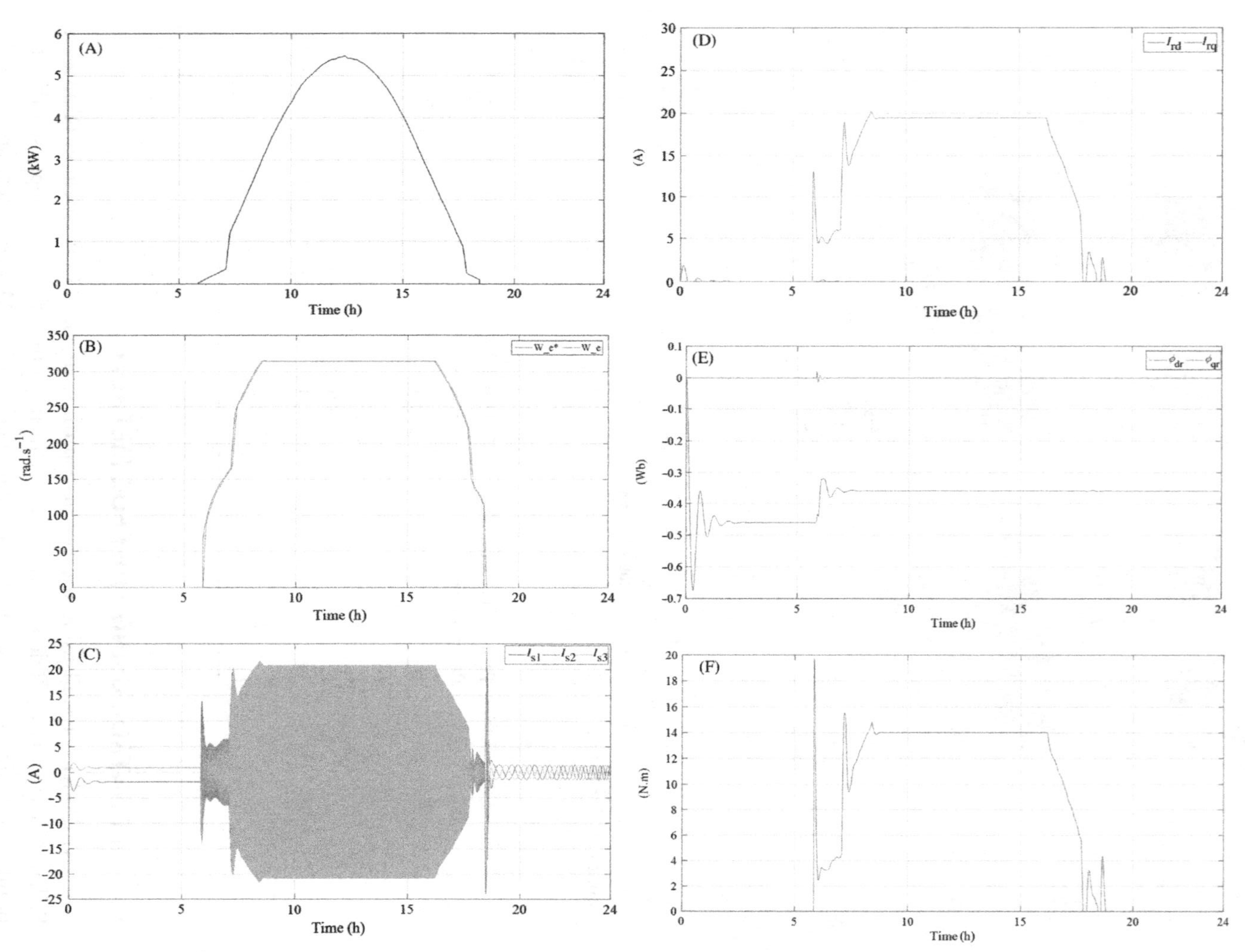

Figure C.4 RFOC results for an IM, (A) P_{pv}; (B) w_m; (C) I_s; (D) I_r; (E) φ_r; (F) C_{em}.

REFERENCES

Kamoun, M. B. A. (2011). *Modeling of electric machines*. Tunisia: Cours at the National School of Engineering of Sfax.

Sallem, S. (2009). Contribution à l'intégration d'une machine doublement alimentée dans des systèmes autonomies. Thesis presented at the National School of Engineering of Sfax, Tunisia.

Yahyaoui, I, Sallem, S, Chaabene, M, & Tadeo, F. (2012). Vector control of an induction motor for photovoltaic pumping. In the proceedings of the International Renewable Energy Conference (IREC), 877–883.

INDEX

Note: Page numbers followed by "*f*" and "*t*" refer to figures and tables, respectively.

N

O

P

Printed in the United States
By Bookmasters